Air Pollution, Climate and Health

Air Pollution, Climate and Health

Edited by:

Stephanya Lynn JonasLabee

Delve
PUBLISHING

www.delvepublishing.com

Air Pollution, Climate and Health

Stephanya Lynn JonasLabee

Delve Publishing
224 Shoreacres Road
Burlington, ON L7L 2H2
Canada
www.delvepublishing.com
Email: orders@arclereducation.com

ISBN: 978-1-77469-108-3 (Hardcover)

ABOUT THE EDITOR

Stephanya Jonas-Labee has been passionate about the environment her whole life. After completing her bachelor's at UCSC, she moved to Amsterdam to earn her Master's of Earth Science with a focus on environmental management and sustainability from the University of Amsterdam. When not reading about new sustainable technologies, she enjoys traveling the world, trying out new vegan recipes, and working on her garden.

TABLE OF CONTENTS

LIST OF FIGURES

Figure 4.4. Air sulfur dioxide is 99% derived from human sources. It is an acidic gas which acts in two ways as a pollutant. First, direct to the lungs and airways of humans in which they are irritating and can cause wheezing, density of the chest and pulmonary disease. For individuals who have asthma and some lung diseases, this can be a special problem. The environmental issue of acid rain is also caused by sulfur dioxide (together with various nitrogen oxides)

Figure 4.5. Researchers found a new chemical reactivity that would serve to further decrease smog-causing nitrogen oxides in diesel exhalation by improving catalyst designs for contaminant control systems

Figure 4.6. Ozone is an anti-UV shield around our planet to protect us from high energy rays, which can, when left unimpeded, penetrate the atmosphere, and cause more skin cancer and cataracts, resulting in further degradation, which is harmful to plants and animals

Figure 4.7. Acrolein is a chemical that is employed in the chemical industry as a transitional reactive aldehyde. It is used to synthesize many organic compounds, the formation of methionine and methyl chloride. Acrolein is exposed to smoking, second-hand smoke, wood, and plastic smoke exposure. Firefighters and people in high motor traffic areas may be exposed to higher levels of acrolein through smoke inhalation or automotive exhaust

Figure 4.8. Most people feel it has been completely prohibited by the number of legal actions and the general upheaval over asbestos. However, there are different regulating levels around the globe for this mineral. The US continues to allow asbestos in products to be contained in 1% or less while Australia has completely banned it

Figure 4.9. Several problems can result in benzene in the air. In the disease control and prevention centers, for example, long-term exposure to elevated benzene levels can produce leukemia

Figure 6.1. Air pollution is mostly replaced with alternative energy sources such as solar, wind, and geothermal fuels. Fossil fuels are the primary solution. It is essential to produce clean energy. But our energy consumption must also be reduced by using responsible practices and more efficient equipment

Figure 6.2. The effects of air pollution are still evident, although laws and regulations contribute. The decrease in toxic air and improvements in health are welcome changes. But there remains a challenge which needs to be addressed by the growing danger of climate change because of emissions of fossil fuel

Figure 6.3. There is a need to request countries to enforce strict regulations requiring air scrubbers in all industrial installations that release pollutants from the atmosphere

Figure 6.4. Pollution is a big term now used by our ears. Every day, we hear about and read through the mass media about the various forms of pollution. Air pollution is a form of air pollution, regardless of indoor or outdoor air contamination

Figure 6.5. Governments, councils, and planning bodies now have to begin to include measurements of air quality in their strategic plans and planning decisions. It is probably

Figure 9.7. Many actions that encourage ecosystem adaptation also help people adapt to ecosystems. For example, restoring natural fire democracies decreases the possibility of catastrophic fires and decrease environmental exposure to this danger. Giving more room to the water bodies enables more water to be stored in the natural system and less probable to flood in populated areas. The supply of green areas and trees gives shade to animals. Trade is made with the reconstruction of ecosystems in certain areas between agricultural production

LIST OF TABLES

LIST OF ABBREVIATIONS

AIDS	Acquired Immunodeficiency Syndrome
BBIs	Blood Biomarkers
CFCs	Chlorofluorocarbons
FAO	Food Agricultural Organization
FGM	Female Genital Mutilation
GHGs	Greenhouse Gases
HAP	Household Air Pollution
HCFCs	Hydrochlorofluorocarbons
HFCs	Hydrofluorocarbons
HPV	Human Papilloma Virus
ICAR	Indian Council of Agricultural Research
IEA	International Energy Agency
LED	Light Emitting Diode
LiST	Live Saved Tool
LPG	Liquefied Petroleum Gas
N_2O_5	Dinitrogen Pentoxide
NAP	National Adaptation Plan
NARP	National Agricultural Research Project
NERC	Natural Environment Research Council
NH_3	Ammonia
NMVOCs	Non-Methane VOCs
NO	Nitric Oxide
NO_2	Nitrogen Dioxide
NOx	Nitrogen Oxides
PAN	Peroxyacetyl Nitrate
Pb	Lead
PG	Piped Natural Gas
PM	Particulate Matter

RCTs	Randomized Controlled Trials
SO_2	Sulfur Dioxide
VOC	Volatile Organic Compounds

ABSTRACT

The climate can make the air less favorable to breathe, thereby affecting human health. Higher temperatures cause an increase in allergens and toxic air particles. The book mentions that the climate can affect human well-being by expanding ground-level ozone or potential particulate air pollution in certain areas. Ground-level ozone (an element of exhaust gas) is associated with many medical problems, including lung malfunction and asthma. The book suggests that contact with allergens makes some people uncomfortable. When they are exposed to allergens and air pollution, allergic reactions tend to become more serious. The increase in toxins in the air makes the impact of related allergens found in the environment more frightening. Individuals who are sensitive to pollen dust may be at increased risk of intense respiratory reactions.

PREFACE

The climate impacts human health in many ways. Extreme changes in weather and environment exacerbate existing health problems and can cause new health problems. This book examines the main risks that climate factors pose to human health and discusses who are most affected by it. Smoke exposure can increase the chances of contracting acute (or sudden onset) respiratory diseases, and hospitalizations for respiratory and cardiovascular diseases.. As drought conditions become more common the rate of wildfires is expected to increase, which will affect air quality in many countries across the globe.

The climate may lead to an increase in insect-borne infections and water-borne syndromes. This is because weather patterns may increase the length of the season for insects to spread infections. Countries may face increased risks of current water-borne and insect-borne diseases as well as diseases that have not yet appeared in the region. Changes in rainfall patterns may also increase the risk of water-borne and infectious diseases that cause diarrhea. Increased temperature can cause or exacerbate a series of serious health problems. Prolonged exposure to extreme heat can cause:

- Heart disease;
- Heatstroke;
- Muscle cramps;
- Exacerbate existing conditions, such as respiratory diseases and heart disease;
- Death.

This book discusses how extreme variations in temperature may have a greater impact on some groups of people, especially children and the elderly. Increased air pollution may pose a greater health risk to these high-risk groups. Higher amounts of dust, ozone, and fine air particles will reduce air quality and cause or aggravate a range of health problems, such as asthma, cough and inflammation, lung inflammation, lung cancer risk, and airway congestion.

Climate change poses a major risk to human health in various ways. Natural disasters and extreme weather events can cause injuries, mental health problems, heightened risk of infection, and death. Damage to health infrastructure such as hospitals can also cause impediments in health and well-being. Environmental pollution, increased toxins, and allergens can all cause increased respiratory and heart health problems. Children, the elderly, and people in economically underdeveloped areas, are most vulnerable to health problems related to climate change. Maintaining a healthy climate is essential to help reduce these impacts and health risks.

Changes in weather affect crops and food production. Food costs may rise, which may lead people to adopt unhealthy diets. A poor diet can lead to hunger, malnutrition, or obesity. Depending on the quantity, the nutritional value of some foods may be reduced because of climate change. The expected increase in carbon dioxide content in the atmosphere and the changes in soil nutrients will lead to the reduction of nutrients in different crops, thereby affecting health and nutrition.

Chapter 1

What is Air Pollution?

CONTENTS

1.1. INTRODUCTION

Air pollution is quite an old concept in this world. Jacobson (2012) claim that air pollution was experienced millions of years ago when a meteorite hit the earth's surface and managed to eliminate dinosaurs. The crash of the meteorite led to the formation of clouds and smoke and a mass extinction event where the climate was changed. To understand air pollution, one needs to understand exactly what air pollution is. Air pollution can be defined as the presence of substances in the atmosphere that are harmful to living organisms, including humans, and may cause damage to the climate or other materials in the world. Air pollutants are not limited to one but rather three types (Zhang, 2017). The first is gases, which include carbon monoxide (CO), sulfur dioxide (SO_2), ammonia (NH3), chlorofluorocarbons (CFCs), nitrous oxides (N20), and methane (CH4). The second type is particulates, which include both organic and inorganic particles the third type involves the biological molecules. Air pollution is very harmful to living organisms as it may lead to allergies, diseases, and even death. Damage to the natural and the built environment may be brought about by air pollution. Both the human activities and other natural processes can generate air pollution (Figure 1.1).

Figure 1.1: Air pollution involves the presence of substances in the atmosphere that are harmful to living organisms.

Source: https://elemental.medium.com/a-surprise-surge-in-air-pollution-may-be-causing-more-coronavirus-complications-c444e82a659e.

Air pollution is considered a risk factor because it is responsible for some of the deadliest diseases such as heart disease, respiratory infections, lung cancer, COPD, and stroke (Kenis & Mathijs, 2012). Poor air quality has damaging effects on an individual's health. Normally, the effect is normally on the respiratory system and the cardiovascular system. Different individuals react differently to air pollution depending on the type of pollution that the person has been exposed to, the person's health status, the degree of exposure, and genetics. In 2008, a report done by the Blacksmith Institute on the World's Worst Polluted Places stated that two of the world's worst toxic pollution problems are indoor air pollution and poor urban air quality. Annually, outdoor air pollution causes millions of deaths worldwide. When considering air pollution in general, the death toll is quite higher. As such, (Jang & Hart (2015) has taken the initiative to name this as the world's largest single environmental risk. Trillions of dollars' worth of the world's economy are lost each year because of productivity losses and poor quality of life brought about by air pollution (An et al., 2018).

Materials that are found in the air that have adverse effects on humans and the ecosystem at large are what are known as air pollutants. They can be in the form of liquids or gases. As described before, pollutants can be natural or manmade. In some cases, air pollution may be caused by natural sources such as volcanic eruptions and forest fires. They tend to release greenhouse gases (GHGs) and other pollutants. Anthropogenic air pollutants such as extreme weather patterns and environmental conditions are typically released to the environment or rather the atmosphere.. Some of the climatic conditions known to cause major pollution include acid rain, ozone depletion, and global warming, which are seen to be persistent as the years go by. When it comes to dealing with fabricated sources of air pollution, they are classified into different categories. The categories can be by source, either nonpoint or point, or by formation, primary or secondary (Zaval et al., 2014). Primary pollutants can be either point or nonpoint. Point sources are from individual identifiable air, thermal, noise and light pollution and cover a defined space. On the other hand, nonpoint sources release pollutants across a wide area. The chemical reaction of the primary pollutants is what leads to secondary pollutants. Their origin is normally from the nonpoint sources. Not only natural factors can lead to air pollution, but there are manmade sources also. Such manmade compounds include fumes and gases from plastics, pesticides, fibers, chemical gases, and heavy metals. Ground-level ozone is an example of a secondary pollutant, which forms part of pollutants that make up photochemical smog.

1.1.1. Major Primary Air Pollutants

1. **Nitrogen Oxides:** This category involves mainly nitrogen dioxides. Nitrogen dioxide (NO_2) is considered as one of the prominent air pollutants. It is a toxic gas with a sharp biting odor and is reddish-brown in color (Figure 1.2). It is normally formed from high combustion burning and can be produced when there are thunderstorms through electric discharge. To observe this, one might recognize it as the plume downwind of sites or as the brown haze dome up in the sky. Nitrogen dioxide forms part of several nitrogen oxides (NOx). (Yeager et al., 2011).

Figure 1.2: Brown haze dome shows the presence of nitrogen dioxide.

Source: https://www.republicworld.com/world-news/us-news/wildfire-smoke-in-us-exposes-millions-to-hazardous-pollution.html.

2. **Carbon Monoxide:** This is a very poisonous gas that is colorless, odorless, and non-irritating. This gas is normally brought about due to incomplete combustion of fuel including wood, natural gas, and coal. One of the major sources of carbon monoxide includes vehicular exhaust. Burning charcoal is another common example of where carbon monoxide is released, which has killed many people. This gas is called a silent killer since it kills swiftly and its effects are hard to detect. (Kellstedt et al., 2008).
3. **Sulfur Oxides:** One chemical in the family of sulfur oxides is sulfur dioxide (SO_2). It is produced in various industrial processes and from volcanoes. Petroleum and coal normally contain sulfur

compounds, and therefore, when they are burnt or combusted they produce SO_2 (Wang, 2015). When SO_2 is further oxidized in the presence of nitrogen dioxide, it forms an acid H_2SO_4 that leads to the formation of acid rain. This is a concern for the environmental economists because these fuels are used as power source, and yet they are harmful to the environment.

4. **Volatile Organic Compounds (VOC):**These compounds are normally subdivided into methane and non-methane. Methane is a vital and efficient GHG that contributes to global warming in an enhanced manner. There are other hydrocarbon VOCs that are important GHGs, especially in their role in the creation of ozone and of prolonging methane in the atmosphere. Depending on the local air quality, the effects of VOCs are different. Non-methane VOCs (NMVOCs) include aromatic compounds such as xylene, benzene, and toluene. These are said to be carcinogens that may cause leukemia if an individual is exposed for longer periods. A dangerous compound of VOC is the 1, 3-butadiene that is often associated with industrial uses (Kgabi et al., 2014).

5. **Particulates:** Particulates, or particle matter (PM) are fine particles or atmospheric particulate matter, which is normally composed of small or rather tiny particles of liquids or solids that are suspended in gas. For instance, aerosol is identified as both a gas and particles. Particulates can be sourced from natural or manmade sources. Some of the particulates that occur naturally are dust storms, volcanoes, living vegetation, forest and grassland fires, and sea spray. Aerosols can also be generated from human activities such as burning fossil fuels in power plants, various industrial processes, and vehicles. Anthropogenic aerosols are those that are produced because of human activities. Scientists claim that over the universe, they account for 10% of the total amount of aerosols in the atmosphere. Health hazards such as heart disease, lung cancer, and altered lung function are linked to the increased levels of fine particles in the world. This is something that is becoming difficult to deal with, especially in third world countries, due to the problems of rough roads and the need to plow land in order to plant. Agriculture forms a major part of their economies and thus makes it hard to avoid such circumstances.

6. **Persistent Free Radicals:** These involve the airborne-fine particles that are found in the atmosphere. They form part of the particulates. These tend to be harmful to the human body as they lead to the cardiopulmonary disease (Hiller, 2011).
7. **Ammonia:** Ammonia (NH_3) is quite common in the atmosphere. Pee from animals and humans contains ammonia, so it is necessary to explore ways to reduce the amounts in the environment. NH_3 can also be emitted from agricultural processes, commonly in the fertilizers that are used to improve the nature of the soil (Al-Amin et al., 2013). One of the characteristics of this gas is that it has a pungent odor. They are also used in the synthesis of various pharmaceuticals either directly or indirectly. Ammonia is used widely even though it is considered hazardous. Either naturally or synthetically produced ammonia is increasingly being used around the world.
8. **Radioactive Pollutants:** These are normally produced because of nuclear explosives, war explosives, nuclear events, and natural processes. Radioactive decay of radon is one of the natural process that brings about radioactive pollution (Pleijel, 2009).
9. **Chlorofluorocarbons (CFCs):** This gas is very harmful to the ozone layer. They are normally emitted from products that have in recent times been banned from use, such as pesticides (Pleijel, 2009).
10. **Persistent Organic Pollutants:** These compounds tend to persist in the environment and bioaccumulate within the food chain. They are normally found in chemicals that are known to travel by air for long ranges. For this reason, they may even be found in countries that had no responsibility whatsoever in producing these compounds. Industrial products are their main sources, which include pesticides such as DDT, Polychlorinated biphenyl, burning of waste products such as dioxins and furans, and industrial by-products (Villar and Krosnick, 2011).

1.1.2. Major Secondary Air Pollutants

1. **Ground Level Ozone:** It is normally formed from nitrogen oxide and VOCs. The troposphere is mainly composed of water vapor and aerosols Ozone forms part of some of the areas of the stratosphere and is known as the ozone layer. Aerosols are

chemical reactions and photochemical that tend to drive most of the chemical processes that take place in the atmosphere every time. Human activities such as increased combustion of fossil fuels are pollutants and high constituents of smog which lead to ozone layer formation in the atmosphere which is harmful (Figure 1.3).

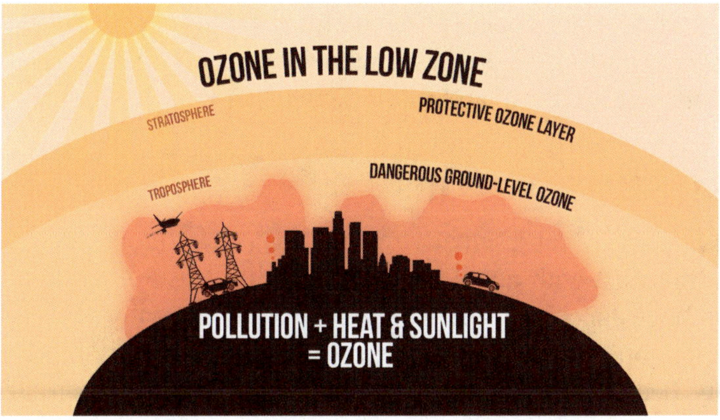

Figure 1.3: Formation of ground-level ozone formed when there is pollution and is exposed by ultraviolet rays.

Source: Climate Central.

2. **Particulates:** While particulates can be considered a primary air pollutant, when the sun's ultraviolet rays interacts with some of these pollutants, they undergo a chemical reaction and become more hazardous. Smog is just a word used to mean smoke and fog. Coal-burning in large amounts in an area and further mixed with SO_2 leads to the classic formation of smog. In the current world, smog is mainly derived from vehicular and industrial emissions. The ultraviolet rays from the sun then act upon them. This leads to the formation of the secondary pollutant that then combine with the primary emissions, which then form a photochemical smog.

The industrial revolution was seen to be a form of breakthrough in terms of air pollution, which was exactly what the world needed to avoid. It has brought about extreme problems to the world, starting with the depletion of the ozone layer. The world is however trying its best to curb this problem and currently there are technologies invented and strategies put in place with the aim of controlling air pollution. Some industrialist and motor transports have taken an initiative to use various types of filters, collectors, and precipitators to prevent air pollution. In most parts of the world, technologies for better fuels and engines have been invented (Bassu et al., 2014). Take, for instance, Tesla Company, which is inventing a car that uses only batteries hence low emissions and underground tunnel roads for cars to reduce pollution from exhaust. In the current society, the emerging trend or rather concept is of zero-emission solar and electric engines, which act as a remedy for excessive air pollution for the current and future time.

1.2. EMISSION FACTORS

Emission factors are representative values that seek to relate the amount of pollutants released with activities related with release of these pollutants. Emissions have for a very long time been vital tools in the development of a state, nation, region, and the world at large. The emission factor data has mostly been useful in the development of inventories on air quality, which help the relevant stakeholders make appropriate decisions in the development of emission and control strategies. The factors are normally represented in terms of distance, volume, weight, and the duration of time taken in the emission of the pollutant. Such factors help in the estimation of amounts of pollutions that can be emitted from various sources (Figure 1.4). In recent times, individuals from the federal sector, state and local governments, and tribal agencies, among others, are applying emission factors in the determination of an emissions site and granting of permits to individuals. These stakeholders have sought the knowledge of scientists and other stakeholders on how to apply emission factors when performing certain activities, especially in the development of permits that are practical in their enforcement (Betts et al., 2011).

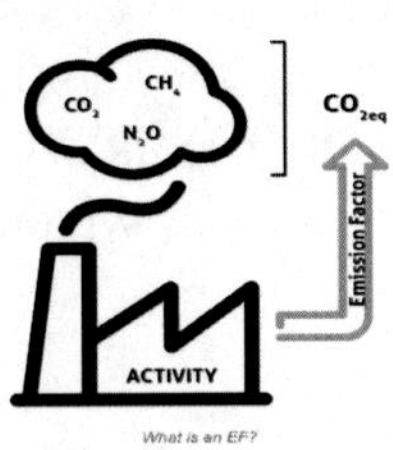

Figure 1.4: Factory pollutants are a leading cause of CO2 pollutants.

Source: https://www.climfoot-project.eu

In the class of persistent organic pollutants, there are 12 compounds. Among them are dioxins and furans, which are intentionally created by the combustion of organics. These compounds have the properties like those of open burning plastics. These compounds also have the properties of endocrine disruptors, also described as hormonally active agents. This is because these chemicals have the ability to disrupt the endocrine system. In short, they have the ability to disrupt the human genetic system. These chemicals have serious medical effects such as birth defects, cancerous tumors, and other developmental disorders, and interfere with the action and elimination of hormones in the body. This then leads to abnormal cell metabolism, which leads to the development of the disorders (Benjamin et al., 2017). There are some groups of individuals that have called for swift action for the removal of endocrine disruptors from the market. However, the process can be challenging given that they are now found in the air given their nature. They form part of the air pollutants and thus becoming quite difficult to get rid of them completely.

There has been a compilation of publication by the United States Environmental Agency on air pollution emission factors that are from a wide range of industrial sources (Schellnhuber, 2008). This is helping the American government understand the amounts of pollutants and therefore having the ability to come up with mitigation measures when it comes to dealing with air pollution. It is not only in America that researchers have taken the initiative to do the same, but this is the case in the United Kingdom, Australia, the European Environmental Agency, and Canada.

1.3. SOURCES OF AIR POLLUTANTS

The various factors of air pollution tend to vary from country to country. In South Africa the main source of air pollution is carbon monoxide from household burning and SO_2 that originate from coal power plants. On the other hand, in Sri Lanka, their main source of air pollution is carbon monoxide (Revesz et al., 2014).

1. **Transport:** This is common source of air pollution in countries all over the world. This is because individuals need transportation to move from place to place and as such pollution results from this movement Transportation is among the leading sources of carbon monoxide. Engines in transportation tools are normally fueled by kerosene, diesel gas, and petrol, which normally combust to allow the locomotive to move. Long-range aircrafts

and jet engines are the main source of NOx (Bose, 2010). When there is traffic on the road, it is normally considered as a nonpoint source of air pollution or line source. In addition to this, turbine engines of huge ships and harbors are considered as sources of toxic particles in the air and GHGs, they cause about 18-30% of the global nitrogen oxide (NOx) and 9% of Sulphur oxide (SOx) pollution (Revesz et al., 2014). Figure 1.5).

Figure 1.5: Transport is a source of air pollution that is common all around the world.

Source: Greenhouse PR.

2. **Sources of Energy Generations:**Pollution occurs in the process of generating energy. It occurs when large amounts of natural gas, biofuels, coal, gasoline, and oil are used in combustion to produce energy. In the process of combustion, carbon oxide and sulfur oxide are produced which are substances harmful to the environment.

3. **Households:** Individuals in the household normally do not have a clue of their effects on the environment. When an individual in the household uses fossil fuels in cooking, there is the production of carbon and soot, which are air pollutants. Insecticides intoxicate the air that we breathe and even the food we eat which then leads to poisoning. In simple terms, one should sit and think, in most cases, it is advised that insecticides be used in a room where food is not stored or cooked. Once an insecticide is used, it is advised that individuals vacate the room to allow fresh air to come in and

to prevent the individuals from breathing in the pollutants. The aerosol sprays that are used in our homesteads for freshening the air are also quite harmful to the environment and human health. They lead to allergies, which are health complications (Buell, 2014).

4. **Industry:** For the industry to function, there need to be emissions to the environment. This is because most industries depend largely on fossil fuels for them to run. Using fossil fuels requires combustion, which leads to the production of carbon monoxide and carbon dioxide. Other compounds that are produced from industries include particle maters and sulfur hexafluoride. In cement industries, for instance, there is a high release of particle matter. This is because there is high involvement of grinding and combustion in order for cement to be produced. There is also a wide range of chemicals that are produced in the manufacturing of electronics, paint, chemical thinning agents, and dry cleaning. The industrial processing of agricultural products also leads to the production of chemicals that are harmful to the body. In tea industries there is a high production of dust particles that are quite tiny and can easily enter the human system (Kellstedt et al., 2008). Furthermore, when industries utilize oxides from nitrogen, SF_6, PFC, and HFC, there will be production of pollutants (Figure 1.6).

Figure 1.6: Most industries depend on fossil fuels for operation but they are a massive source of air pollution.

Source: https://www.theguardian.com/environment/2013/apr/17/why-cant-we-give-up-fossil-fuels.

5. **Agricultural Practices:** Common agricultural production also involves the production of pollutants. Starting with tilling of the ground, there are tiny particles that are swept up by the wind which then pollute the environment.. Natural fertilizers have the tendency to produce GHGs (Victor, 2011). In the use of pesticides, there is a release of persistent organic pollutants. In the process of cattle ranching, there is the production of GHGs, mainly methane because of the process of enteric fermentation. In the production of biogas by the use of cow water, there is also the production of GHGs, which are hazardous to the environment (Victor, 2011). Weedicides and pesticides also contain toxic chemicals that tend to have negative impacts on the quality of air that is inhaled. As such, an individual needs to be careful of the farm inputs they use, as they can be very toxic to their health.
6. **Construction and Repair Works:** Dust is normally generated in construction and repair works. Such activities include drilling, blasting, transportation, loading, and offloading of materials, automobile repairing, welding, and painting, among others.
7. **Land Mining, Quarrying, and Other Land Moving Activities:** These activities involve mining large mineral deposits from the earth's ground. It also includes blasting and quarrying of limestone for the case of cement manufactures.
8. **Burning of Waste and Incinerators:** Waste is burned daily in various settings: outside our homes, in industrial processes, and even in the hospitals. In fact, hospitals tend to burn some of the most toxic plastics and chemical substances. Researchers and scientists have made claims that this might even be the most severe threat to the environment (Victor, 2011). This is because it contaminates the environment with persistent organic pollutants, which originate mainly from burning hazardous waste like soiled surgery implements, bandages, etc.. This is in addition to the common compounds that are produced during combustion, such as carbon oxides and soot. Waste presents itself in a number of ways, including biological and chemical waste, industrial chemicals, electronic waste, drugs, glass, steel, cement dust, plastic, and various soil mineral derivatives(Trenberth et al., 2014).

9. **Natural Sources:** Nature also has its ways of emitting pollutants into the environment. For example, with volcanic activities there is the production of compounds such as ash, black smoke, metals, carbon oxides and sulfur oxides. In other natural processes, there is the production of methane primarily from thawing of the permafrost regions in the northern hemisphere, wetlands, and sanitary landfills. Other examples include dust storms, forest and bush fires, conversion of land use, and the release of terpenes and isoprene from forests and sea sprays (Figure 1.7).

Figure 1.7: Volcanic activities lead to the production of ash, metals, black smoke, among other compounds.

Source: Britannica.

1.4. GENERAL EFFECTS OF AIR POLLUTION

One of the major effects of air pollution is health effects. According to the World Health Organization, air pollution has a number of health effects including lung cancer, respiratory infections, and heart disease. Some air pollutants are endocrine disruptors and affect human cell metabolism. Some of the observable effects that are brought about by air pollution include difficulty in breathing, wheezing, coughing, sneezing, asthma, and it further aggravates the some of the already existing cardiac conditions. As a result, an individual who is suffering from such a situation is forced to increase their medication use, which may be harmful and can be quite expensive for an individual to handle. To make matters worse, it can lead to increased emergency and doctors' visits, individuals being frequently admitted to the hospital, and even premature death. The effects of air pollution on the health of a human being are much more, but the effects on the respiratory system and the cardiovascular system are more common. How an individual reacts to air pollution varies mainly considering the type of pollutant that the individual has been exposed to (Brulle et al., 2012).

Some of the common sources of air pollution as mentioned include sulfur oxide, nitrous oxide, and ozone. The number of deaths that have been caused by both indoor and outdoor air pollution is far more reaching and quite alarming. In a single year the number of deaths caused by air pollution alone is about 3.3 million (Pleijel, 2009). The developing countries are currently trying to meet the standards of the developed countries in terms of industrialization. As such, industries have been built, but the problem is that the health sector has not yet been improved, leading to a high number of deaths in these countries. Children below the age of five are the ones that are falling victims of these circumstances..

Bhopal Disaster that occurred in India in 1984 was one of the worst short-term pollution crises that has occurred in the world. This led to the death of about 15,000 to 20,000civilians in the country, and since then, some mitigation measures have been put in place in order to avoid such situations. In the USA, 25,000 individuals died, and between 150,000 and 600,000 of the people were outright or injured because of the leaked industrial vapors that came from the Union Carbide factory that belongs to Union Carbide. In the United Kingdom the worst air pollution case was experienced when a Great Smog formed over London in December 1952. Within six days, a huge tragedy was experienced, with over 4,000 deaths and over 8,000 died over the coming months. This was a huge tragedy in the country. In 1979, it is said that there was an accidental leak of the anthrax spores near Sverdlovsk that originated from a biological warfare laboratory and led to the deaths of hundreds of civilians (Singh and Purohit, 2014).

In the world, individuals, especially children that are living in areas that are exposed to air pollutants, have a risk of developing terminal respiratory conditions such as asthma, pneumonia, and other lower respiratory infections.. Industries tend to create more job opportunities for civilians, and further with their existence, the infrastructure within the area is developed. Children in such areas tend to be at higher risks because they are exposed more as compared to the adults. Children tend to go out more to play and thus are exposed to what is known as higher minute ventilation and thus at more risk to the dangers of air pollution. Pregnant mothers are also highly affected, which then leads to their babies having low initial birth weight (Pleijel, 2009).

The environment is at a high risk of destruction if individuals continue to not give much attention to what they are doing. Poisonous air pollutantslead to the formation of acid rain. In this case, plants, animals, and human beings are at a high risk of being harmed when acid rain falls on the plants, they are withered because of the high ph. In this case scenario, individuals are forced to plant their crops in greenhouses to avoid hunger in the coming years. This shows that with air pollution, there is the destruction of crops, animals, farms, and human beings. In addition to this, acid rain makes water bodies acidic, thus making it harmful for animals and humans that depend on the water.

Air pollution has an impact on the world's economy. In this case, the effect is much indirect as compared to their effects on the environment and the health of an individual. This is what researcher's term as a secondary effect. A healthy population results in an increase in productivity which creates a healthy economic system." (D'Amato and Akdis, 2020). Take, for instance, the old generation are needed to teach the young generation the ropes of the economy. The young people are on the other hand are very important as they are the future of the society. Businesses depend on raw materials for finished products to be created, and this is where the farmers chip in. They also require natural resources for them to fully run efficiently. As air pollution destroys the environment, crops are destroyed. Among these crops, some of them are food products, and when food is not available, hunger strikes and individuals die. Some of the other crops are used as raw materials to create finished products to run the world. With a lack of raw materials, some of the industries will lose their ability to operate and shut down. In this case, people will lose jobs and cause a fall in the economy at large..

It is with no doubt that air pollution has adverse effects and if the individuals around the world are not cautious enough, this type of pollution can cause death around the world. In fact, it may lead to the extinction of humans in the world. Instead of scientists trying to discover new planets and ways for humans to survive off the face of the earth, it will be important for them to try to come up with measures that are useful in mitigating air pollution (Jang & Hart, 2015). Furthermore, companies such as Tesla need to be promoted in order to encourage the use of energy sources that do not depend entirely on the combustion of fossil fuels. In fact, what individuals should try to does to try to embrace their past During pre-industrial times, air pollution was not considered to be a big issue since it were only caused by natural events rather than there being manmade sources of air pollution. Furthermore, individuals need to find measures to avoid air pollution entirely. Government states need to come up with rules and regulations to protect their countries from the effects of air pollution.

1.5. MEASURE OF PREVENTING AIR POLLUTION

Given that air pollution dulls the sunlight and obscures views, damages the environment, and contributes to major health problems, the need for measures that mitigate air pollution cannot be overemphasized. Reduction of pollution has a positive effect on health, particularly the health of susceptible individuals.. Levels of air pollution can be reduced using various strategies with the government playing a key role (Dai, 2011). For example, during the 2008 Olympic Games, the Chinese government was able to implement measures that controlled air pollution. This resulted in a decline of at least half of the average number of outpatients visit from asthma during the Olympics period as compared to before the start of the game (Jang & Hart, 2015). In another study of Beijing, children showed that fractional exhaled nitric oxide (NO) levels were significantly lower during the Olympics period as compared to before and after Olympics, suggesting that rapid inflammatory changes occurred (Dai, 2011).

Properly vented wood burning cooking stoves versus open fires showed reductions in the longitudinal decline in forced expiratory volume and improvements in respiratory symptoms when proper cooking stoves were used in rural Mexico. The use of improved cooking stoves has also been found to reduce 50% of carbon monoxide emissions and resulted in a lower rate of diagnosis of pneumonia. Evidence is accumulating that polymorphisms in several genes involved in oxidative stress play an important role in susceptibility (Robinson et al., 2009). These genes include those coding for phase II enzymes, all of which mediate the detoxification and elimination of toxic products. GSTM1 and GSTP1 have been the most frequently studied enzymes. A study of Mexican children exposed to high concentrations of O_3

reported that an association between polymorphisms in the oxidative stress-related GSTM1 gene and the development of asthma. Interestingly, children with the GSTM1-null genotype were more susceptible to the effects of O_3. Taking supplements containing the antioxidant vitamins C and E conferred protection against O_3 exposure (Robinson et al., 2009).

Other studies have reported on measures such as taxes, grants, licensing, and voluntary schemes to control and reduce emissions from industry. The Belarusian National Action Plan on Rational Use of Natural Resources and Environment for 2006–2010 included the introduction of automatic monitoring systems for pollutants and emission sources on the edge of buffer zones and the implementation of environmental management systems. Belarus also reported using basic economic mechanisms, including licensing and taxes. A 10% tax relief was granted to companies with certified environmental management systems, acting as an incentive. In the Czech Republic, a system of charges applied to polluters, according to the operator's size, and the revenue generated entered the State environmental fund for use in environmental projects. Cyprus study reported on the use of incentives such as cash grants to reduce industrial pollution and to promote renewable energy, while Slovenia's public Environmental Fund provided soft loans to companies for activities related to environmental protection, including air pollution prevention and control (Hiltner, 2009). Slovenia also provided subsidies for measures to improve energy efficiency and to support the use of renewable energy. It was noted that the licenses granted to operators in the industrial sector outlined their obligations and guidelines related to efficient use of energy and raw material, optimizing the production processes, avoiding waste, and promoting recycling, etc. It also reported that it ran a "help desk" for companies under its Eco Innovations scheme to provide advice and support to businesses wishing to apply eco-efficient technologies. Further efforts in research reported that efforts to manage energy demand by monitoring energy consumption while raising awareness among industry operators about available options to reduce energy consumption were imperative (Diffenbaugh and Burke, 2019).

On specific subsectors such as the chemical and petrochemical industry, the strategy includes applying charges for negative environmental impacts, setting emission limits, and promoting incentives that lead to low air pollution.. The scheme covers over 11,000 energy-intensive installations, including combustion plants, oil refineries, coke ovens, iron, and steel plants, and factories making cement, glass, lime, brick, ceramics, pulp, and paper across the EU, accounting for about half of Europe's CO_2 emissions.

Stakeholders came up with ways of reducing transport emissions which entailed the reduced use of cars and embracing of cycling and use of public transportation, elimination of old polluting cars, promotion of cleaner cars, and use of improved fuels in the vehicles. Many countries and states including Germany, under its 2002 national cycling plan, have promoted cycling over vehicle use. Moreover, Italy set up a Fund for Sustainable Mobility for the period 2007 to 2009 to reduce emissions by promoting better public transportation, notably through underground railways, trams, and trains. Croatia reported on the creation of bicycle lanes, improvements in the quality of public transport and intelligent traffic regulation. Other countries including Poland are promoting public transport through integrated train, bus, and tram tickets and expanding parking areas near train stations. Portugal further prioritizes improvements to the public transport system, notably by expanding the Lisbon and Oporto subway network, improving the national railway service, and enlarging the fleet of vehicles powered by natural gas in public transport in Lisbon and Oporto. In Romania, there was modernizing of rolling stock fleet to promote rail transport. The Russian government further reported on its efforts to modernize the national transport system. Slovenia introduced "Park and Ride" systems to encourage drivers to minimize the distance driven in their cars and to switch to public transport. In Sweden, some local municipality buses were made to be free of charge for all citizens or special groups, such as the elderly.. Nearly all responding parties reported on financial support schemes to promote the renewal of the car fleet. For example, in the Netherlands, owners of old cars received grants to trade them for newer and more eco-friendly vehicles.

Belarus succeeded in switching vehicles to compressed and liquefied gas and renewed its bus stock. It also introduced higher import taxes for vehicles older than 15 years, with a resulting rise in the importation of newer and cleaner vehicles. Many countries also imposed extra taxes on cars that either had more polluting engines (larger engines) or diesel cars. In parallel, many Parties reported on incentives and subsidies to promote hybrid or electric cars. The Portuguese government provided a subsidy of about €5,000 to the first 5,000 individuals that acquired an electric vehicle in the country. Various Parties reported on emission reduction measures targeting fuels, including incentives to promote cleaner fuels such as biofuels or compressed natural gas. In Slovakia, part of the bus fleet was being replaced with vehicles powered by natural gas. Similarly, Slovenia had included in Ljubljana's bus fleet 20 buses operating on pure biodiesel.

In Sweden, all large petrol stations were required to have a biofuel pump and the installation was supported financially by the state (Hiller, 2011).

Some governments reported ongoing efforts in research and development to improve the transport sector's environmental impact. The Netherlands published several tenders for innovative developments, notably one to reduce the number of kilometers necessary for transporting agricultural products, such as food and flowers/plants. One approach trialed in the United Kingdom by the Highways Agency was to examine the effectiveness of a barrier coated with titanium dioxide designed to remove nitrogen dioxide from the air (Donnelly et al., 2017). The Russian Federation reported also on its air and maritime transport policies. In the field of air transport, it implemented improved environmental standards for aircraft engines in line with those of the International Civil Aviation Organization, and was seeking to set high standards in civil shipbuilding for its industry to become more competitive globally. It also set specific measures for ships to reduce their emissions of SO_2, NO_2, and VOCs. With respect to on-road vehicles, the Russian Federation reported on the application of vehicle taxes.

Past studies have placed emphasis on promoting energy efficiency which is not only contributing to reducing air pollution but also reducing anthropogenic sources of GHGs responsible for climate change. In its 2006–2010 energy policy, Belarus promoted the use of alternative and renewable energy, the implementation of combined schemes of power generation, the use of technological measures to suppress the formation of nitrogen dioxide, the modernization of boilers, the development of small power stations, and emission controls. Canada promoted energy efficiency through regulations, incentives, and non-financial assistance for voluntary action, information, and outreach activities. Measures under Croatia's Energy Efficiency Master Plan (2008–2016) included the use of biodegradable municipal wastes to fuel district heating plants, credits for renewable energy projects, and the promotion of energy efficiency in buildings. Cyprus's energy policy focused on energy pricing, sector specific energy-efficiency programs and renewable energy. Combined heat and electricity generation was promoted in the Czech Republic, Finland, and Germany. Some countries such as Germany still continue with these policies up to date.

Italy promoted energy savings and renewable energy. In Slovakia, measures to increase energy efficiency included regular inspections of boilers, heating systems and air-conditioning systems and the labeling of appliances according to their level of energy consumption. In Slovenia, the most important energy efficiency measures were fiscal measures, such as favorable taxation for biofuels for transport; market measures, including the use of certification for energy in buildings and for identifying the origin of electricity. Italy supported energy-efficient buildings through fiscal deductions of up to 55% and municipalities could reduce ownership tax to below 4% if renewable energy systems were installed. Similarly, in the Netherlands, since 1 July 2009, improvements in the energy efficiency of buildings were eligible for tax deductions and several subsidies existed to support energy improvements to private buildings. Liechtenstein promoted energy efficiency in buildings through its Building Act, which notably subsidized installations using renewable energy such as solar energy, wood, and biomass.

Renewable energy sources were a major feature of all responding Parties' approaches to reduce emissions from the energy sector and were generally supported by grants or subsidies. For example, in Bulgaria, renewable energy represented about 8.67% of total energy consumption in 2004, and about 11.49% 1 year later. Croatia aimed for its share of electricity from renewable sources to reach about 5.8% of total electricity consumption in 2010 compared to about 0.8% in 2004, so far they have achieved the goal. A major objective of Cyprus's policy to reduce emissions was to increase the contribution of renewable energy to the energy balance of the island up to 12% by the end of 2010. In the Czech Republic, renewable energy contributed 4.8% of the domestic energy consumption in 2008, essentially hydropower, but also biomass, biogas, and solar energy that had increased significantly (Fountain, 2019). Furthermore, the Czech Republic reported that the government subsidized up to 85% of the total eligible expenditures of a renewable energy project. Denmark's energy strategy aimed at increasing the share of renewable energy to 20% by 2011. To reach this target, the government subsidized onshore and offshore wind power and set up a special fund of DKK 25 million (approximately €3.3 million) per year over 4 years to promote the installation of solar photovoltaic cells, wave power, fuel cells running on renewable fuels and other renewable energy sources. In Finland in 2019, about 31% of electricity produced was from renewable sources, mainly hydropower and biomass.

Hungary increased its use of renewable energy sources, particularly that of biomass, to reduce air pollution. Poland's "Strategy for the development of renewable energy," adopted in 2001, aimed to increase the share of renewable sources of energy, particularly hydropower and biofuels to at least 16% by 2020 through low interest loans and subsidies. Portugal was committed to securing 32% of its share of energy from renewable sources by 2020, with individual sources such as wind and biomass each having a specific target. To support these efforts, it created an "Innovation Support Fund" in 2008 with a total of €76.8 million. Romania reported on the use of a trading scheme whereby mandatory quotas of green certificates from renewable energy sources were allocated and could be traded. In Slovenia, the use of biofuels and other renewable fuels were promoted for motor vehicles, with the aim for them to reach 5% of all transport fuels by 2010 and 7.5% by 2015. More ambitiously, Slovenia's National Energy Program expected that by 2010, the share of renewable energy would increase up to 12% of total primary energy supply, and the share of electricity based on renewable energy was expected to reach 33.6% of final electricity consumption. Spain's 1999 Renewable Energy Promotion Plan committed to achieving 12% of primary energy use from renewable by 2010. Its Renewable Energy Plan for 2005–2010 foresaw that by 2010, 31% of electric production would come from renewable energy, and approximately 7% of transport fuels would be biofuels (Haines et al., 2007). Switzerland's program Swiss Energy was launched in 2000 with the objective of improving energy efficiency and the use of renewable energy, mainly biomass, wind generation, and passive solar systems to help meet the target set out in the Kyoto Protocol to the United Nations Framework Convention on Climate Change. The United Kingdom reported on its "Renewable Obligation," which required licensed electricity suppliers to source a specified and increasing proportion of their electricity from renewable sources or pay a penalty. As a result, the proportion of electricity from renewable sources rose from 1.8% in 2002 to 5.4% in 2008. The United Kingdom also reported on the anticipated economic impact of its investment in offshore wind energy, which is estimated as being worth up to £75 billion (approximately €90 billion) by 2020, and providing up to 70,000 jobs. Measures reported by Parties to address emissions from agriculture included alternative livestock feeding strategies, improved practices for manure storage and spreading, low-emission animal housing systems, and the use of mineral fertilizers.

For several parties, biogas presented an opportunity to reduce emissions from agricultural waste while at the same time improving energy efficiency. Various economic instruments were used by Parties in the agriculture sector, such as subsidies, tax incentives, fees, and low interest loans. For example, tax benefits were granted in the Netherlands for farms that used techniques leading to low emissions of ammonia and GHGs; Norway imposed fees on pesticides; and both Romania and Spain provided incentives to renew agricultural machinery. Many parties promoted organic farming through various financial schemes. In Cyprus, under the Rural Development Programmed 2007 to 2013, cash grants were provided for organic farming. The Czech Republic and Denmark reported using subsidies to promote organic farming in order to reduce air pollution caused by using synthetic fertilizers for plant growth (Herndon, 2018). A sharp rise in the proportion of organic farming was seen in Slovenia thanks to financial support from the state, with the share of organic farming to total farming area increasing from 0.5% in 1999 to 6.1% in 2008. Switzerland provided direct payments for organic farming. Some Parties reported on activities related to awareness-raising and information dissemination on organic food to stimulate demand. The United Kingdom reported on a scheme called the Organic Conversion Information Service, which targeted farmers and provided them with free advice on the process and implications of switching to organic farming. This contributed to the protection of natural resources, biodiversity, and animal welfare, as well as to rural development. Many parties mentioned that their research activities were linked to those of the Convention's Cooperative Program on Monitoring and Evaluation of the Long-range Transmission of Air Pollutants in Europe and its task forces. The Czech Republic reported that in 2010 it would participate in a case study on heavy metals, which would analyze factors affecting the quality of the assessment of heavy metal pollution levels. Slovakia reported an environmental information system accessible online which provided verified information on the environment and served as an important decision support tool.

A number of parties highlighted the importance of their air quality monitoring stations. For example, Hungary had 59 air quality monitoring stations across the country, which measured levels of SO_2, NO_2, carbon monoxide (CO), PM, NMVOCs, and ozone. Similarly, Romania reported that its National Air Quality Monitoring Network had 142 automatic measuring stations across the country, which measured air concentrations of SO_2, NO_2, CO, benzene, and ozone, and PM. The Russian Federation reported that it had monitored air quality through 699 stations in 248 cities in 2008. Switzerland also reported that it had 96 stations to monitor air pollution across the country, 16 at the national level and 80 at the cantonal level. The EU had a number of innovative programs, such as the Competitiveness and Innovation Framework Program's Entrepreneurship and Innovation Programed, which provided direct financial support for new and innovative technologies (Herndon and Whiteside, 2019).

In conclusion, with respect to air pollution legislation and policies, all responding parties reported that they conducted public consultations prior to passing new laws or amending existing ones. The public was invited to participate in the preparation or revision of plans related to air pollution through open public hearings. Germany sent drafts of legislation specifically to selected experts and published them online. In Poland and the United Kingdom, the public was invited to comment on any new Government proposal within a period varying from 21 days to 12 weeks. National and regional authorities, private entities, and non-governmental organizations held workshops, seminars, or conferences to raise public awareness about new policies and to improve the application of existing ones in Portugal. Thus, efforts in implementing the strategies to mitigate air pollution are necessary for all stakeholders, including all public participation.

1.6. CONTROL DEVICES

There are many devices that are commonly used by industries and locomotive devices for the purpose of controlling the level of pollutants that are exposed to the environment. The functioning of these devices involves the destruction of contaminants or their removal from an exhaust stream before they are emitted or rather exposed to the environment.

1. **Mechanical Collectors:** When using this method, dust is separated using centrifugal or gravitational forces. The use of centrifugal forces involves the application of a cyclone. This method tends to protect the fabric filter from coarse particles, and is thus considered an excellent pre-collector. A cyclone that is mechanically driven can collect particles below 10 micros more effectively. There are two types of cyclones, either a multi-cyclone or parallel. A wet cyclone is considered more effective, and this can be achieved by spraying some water to the cyclone (Hiltner, 2009). This allows for the agglomeration of particles and further avoids the re-entrainment of dust.. When the load of dust is quite high, large particles are first removed using this method. A baffle chamber is normally used in the prevention of glowing particles from entering the inner fabric and further prevent it from burning. Apart from this, when the dust is more coarse and abrasive, they tend to be distracted from flowing in order to protect the bag with the fabric filter.

2. **Fabric Filter:** "Filtration size is an important factor to consider when deciding what type of filter to use.". This further determines the mechanism of capture such as diffusion, sieving, electrostatic attraction inertial impaction, and direct interception. The performance of the fabric filter depends on parameters including temperature, chemical composition of dust, dew point and moisture, operation pressure, and distribution of particle size. Most of the fabrics that are meant for filtration are normally made from either polypropylene, polyester, polyphenylene sulfide, polyamides, Fiberglass, Polytetrafluoroethylene, acrylic, cotton wool, Gore-Tex, and ceramic, among other types of materials. The type of weave determines the efficiency of the fabric filter. The filter type may be woven in the form of monofilament, spun yarn, multifilament, and a spun yard or such as a plain weave, sateen, and twill. Finishing of a fabric is also quite important an aspect. The fabric is put through a silicone treatment to give it a smooth surface and further reduce the negative properties that fabrics tend to have. One of these negative properties is heat shrinking, where a fabric shrinks when exposed to heat, abrasion. Apart from this, the treatment help in the antistatic treatment, calendaring, which involves the smoothening of the fiber in order to ensure easy dust release. A coated finishing of Teflon is also applied, which improves the efficiency of the fabric, and enhances the cake release. An antistatic treatment is also done on the fabric to allow it to handle all of the charged particles and prevent the fabric from being flammable and further avoid explosion in the industry. For the case of non-woven fabric, maintenance is an important aspect. As such, once installed, they need to be regularly checked, cleaned, and further repaired when they are torn. In the case of dry filtrates such as grain, dust, and fly ash, individual industries need to apply air filtration systems.

3. **Electrostatic Precipitator:** At this stage, the dust particles are passed through an electric field and then charged to saturation levels. Electric forces that cause the charged particles to move in the direction of an electrode are then trapped as they are discarded. This stage has five phases, which include the ionization and corona formation, charging of particles, migration, and precipitation of particles and the removal of dust deposits. In the formation of a negative corona, an electrode is negatively charged and a narrow

negatively charged region called a corona is generated. In this case, the electrons are absorbed by gas molecules. In the space outside the corona there are negative ions and, in this region, the dust particles collide with them and then move towards the positive electrode (Hiller, 2011). In the case of a positive electrode, the process happens in an opposite manner. Negative corona is mostly applied in industries as most of the emission gases can absorb the free electrons the spark overvoltage that is higher which is always so in the case of negative coronas. Air conditioners however, do not apply the use of negative coronas because they generate high levels of ozone, which is harmful for human consumption.

4. **Scrubber:** These are considered as the most efficient types of pollutant removers. In this case, the dust and gas particles are removed using either solid or liquid particles, which flow and intimately get in contact with the gas or dust particles. The mechanism is similar to that of the removal of dust during the rains from the atmosphere. This method is used in foundries, which means they are used in industries of iron and steel. They are also used in the fertilizer industry, in creating nonferrous alloys, and in chemical companies. The advantage of using this method is that micron-level dust can be captured. For the case of liquid scrubbers, there is collection of both gaseous and dust particles (Jang and Hart, 2015). Furthermore, they tend to handle high temperatures and the gases that are said to be corrosive, and it is very difficult to find that they are at risk of catching fires. The generation of a slurry discharge is among the main disadvantages of a liquid scrubber since it leads to water pollution. There is one known as a particle scrubber whose main function is to collect dust particles. Examples of particle scrubbers include centrifugal spray scrubber, self-induced spray or impingement scrubber, spray tower, gas absorption scrubber, and a venturi scrubber. The venturi scrubber is the most efficient particulate scrubber, and is normally recommended for the case of sticky and corrosive dust. The gas absorption scrubber is normally used when the pollutant gas is soluble to the scrubber liquid. In this case, there is the physical transformation of gas molecules to liquid.

5. **Flare and Thermos Oxidizers:** This method is used to burn off flammable gas. They are mostly used in natural gas processing

plants and petroleum refineries. It is further used to release the pressure from the equipment used in the process of controlling air pollution. Flares are normally designed to cater for only short-term combustion. For example, in beer factories, to reduce methane release during fermentation, a flare gas is used.

6. **Conditioning Gas:** The characters of gaseous effluents and the removal of dust particles are normally altered using this process. The major component of conditioning gas is that it works by mixing the ambient air with the effluent. The process dilutes the flow and further reduces the temperatures. This method, however, requires large dust collectors due to the subsequent increase in the mass of air.
7. Other methods used to control air pollution include the use of the local exhaust ventilation systems, fans, and chimneys. However, these methods are only efficient when trying to reduce indoor air pollution. They are not quite effective when dealing with outdoor air pollution.

Air pollution is a great challenge and should be reduced to protect our health, the environment, and the economy. There are different types of air pollution, categorized by the source point and the harmful effects they cause. Some of the methods to reduce air pollution include government policies like promoting cycling, rather than the use of vehicles

Chapter

2

What is Climate?

CONTENTS

2.1. INTRODUCTION

The average weather in a place or rather a particular region over many years is climate. Unlike weather that changes within a few hours, climate takes hundreds, thousands, and even millions of years to change (Figure 2.1).

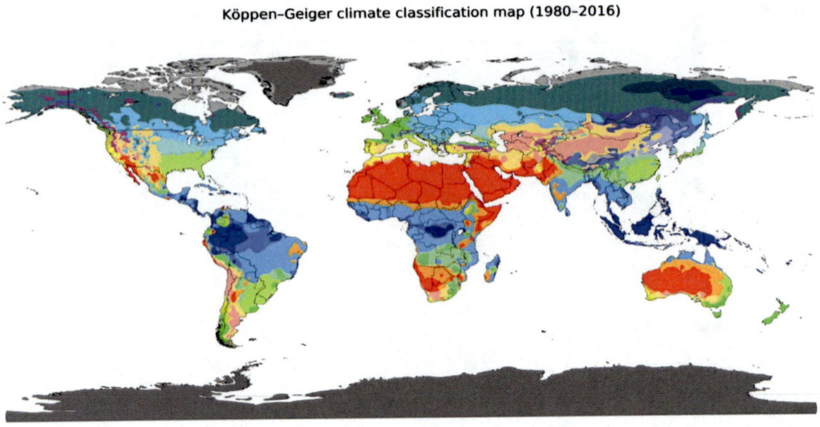

Figure 2.1: The components of the climatic system are what is described as climate.

Source: https://www.nationalgeographic.org/encyclopedia/koppen-climate-classification-system/.

The mean variability of various metrological variables, measured in terms of atmospheric pressure precipitation wind end temperature is denoted by climate. Ideally, the state of the components in the climate system, including ocean and ice on earth is what is described by climate. A location's climate is commonly affected by terrain, altitude latitude and the various water bodies and their current. Ideally, climate is a phenomenon that occurs in our everyday life. It can be classified according to the typical ranges of different variables, the average is and most commonly the precipitation and temperature (Jacobson, 2012). The classification scheme that was commonly used for climate is the Copian climate classification. Since 1948, the Thornthwaite system has been in use. This climate classification system incorporates evapotranspiration along with precipitation information and temperature and is normally used in studying the biological diversity and how climate change affects it. There are, however, other classification systems that focus on the origin of air masses and define the climate of a particular region. These are the Bergeron and the spatial synoptic classification systems (Figure 2.2).

WORLD
METEOROLOGICAL
ORGANIZATION

Figure 2.2: This organization is responsible for describing the climatic trends in the world.

Source: https://public.wmo.int/en/our-mandate.

This study of ancient climates is what is known as paleoclimatology. There are very few direct observations of climatic changes for all times, that is before the 19th century. As such, to acquire data for paleoclimates, scientists infer from proxy variables that include non-biotic evidence such as the sediments found on ice cores and Lake Beds. The proxy variables can also be inferred from biotic evidence such as corals and tree rings. Climatic models are considered to be mathematical models of the past, present, and future climate changes. Just like any change, climate changes may occur either over short timescales or over long timescales. This may be due to other factors, including there is not warming, such as global warming. Redistributions bring about global warming. For example, a small change in the mean annual temperature may lead to a shift Cortana in the isotherms and elevation in the temperate zones (Hiller, 2011)

According to ancient Greek, the term climate was commonly referred to as weather that has been averaged over a long period. Periods may be used to define the climate of a particular place, but the standard period was considered to be 30 years. It is not only the average that constitutes the climate of a particular region but may also include the statistics. Such statistics involve the magnitude of a day-to-day or year-to-year variation. According to the Intergovernmental Panel on Climate Change, climate is described as the average weather, or more rigorously, as the statistical description in terms of the mean and variability of relevant quantities over a period ranging from months to thousands or millions of years. As defined by the World Meteorological Organization, the average or rather the classical period to define climate is 30 years. The quantities that are in most cases considered are normally the wind, precipitation, and most important the temperature of a particular region. In a much broader sense, in order to understand what climate is, one needs to understand the statistical description as well as the climate system at large. According to the World Meteorological Organization, climate normals are used to describe the reference points that are normally used by climatologists in situations whereby they need to compare the current climatic trends with that of the past years (Kgabi et al., 2014). They can also do the comparison with what they consider to be the typical climatic condition. Ideally, climate normals are described as the arithmetic averages of the various climate elements such as wind, precipitation, and temperature. They are normally used in the description of climate over a 30-year period. The 30-year period is considered as an ideal period since within this particular time, one can easily be able to filter out any of the interannual variations or the anomalies that may take place around the world. Furthermore, it is considered to be short enough to show longer climatic trends within the atmosphere.

The origin of the WMO is the International Meteorological Organization that has been in existence since 1929 and was responsible for setting up a technical commission that was left in charge of climatology. A meeting was held in 1934 in Wiesbaden in which the technical commission was responsible for designating the 30-year period as the reference time. At the time, the years ranged from 1901 to 1930 and was from then considered as the reference time to be used by climatologists as the standard normal. In 1982, another meeting was held by the WMO in order to update the climatic condition of the world and the reference time was also 30 years and was from 1961 to 1990 (Zhang, 2017)

In most instances, individuals tend to confuse between climate and weather. In most cases, the situation is that individuals claim that a change in the weather is also described as the change in the climatic conditions. However, there is a popular phrase that is normally used in order to enable individuals to understand the difference between weather and climate. The phrase states that "climate is what one expects while climate is what one gets." Over the past years that even from the period of historic times, there are some variables that have been used to describe the climatic condition which include the latitude, portion of land to water, altitude, and the proximity from ocean to mountains. All these variables that have been stated above tend to change only after millions of years, and it is very hard for an individual to observe the changes. The ideal explanation is that the occurrence of such situations normally outlives a person. Take, for instance, a person is expected to live for approximately 100 years while these changes take place for over millions of years. The climatic changes take longer periods to occur due to a process described as plate tectonics. Some of the other climatic conditions that are considered to be more dynamic include the warming of the northern Atlantic Ocean as compared to other ocean basins and thermohaline circulation of the ocean, which leads to an increase in the temperatures by 5°C. With other ocean currents, there is the redistribution of heat between the land and the water at a regional scale. The effects of solar heat on absorption, water retention, and the rainfall at the regional level normally depends on the type and the density of vegetation in the particular area (Kellstedt et al., 2008). The quantity of solar energy that is retained by the planet that is normally the cause of global warming or cooling is normally brought about by alterations in the atmospheric greenhouse gases (GHGs). The variables that are used in the determination of the climate of the country are numerous, and their interactions are normally very complex. However, there is a general agreement that the broad outlines are normally understood. Furthermore, the determinants of historical climate change are normally put into perspective.

2.2. CLIMATE VARIABILITY

The variations in the mean state and the other characteristics of climate, such as the possibility of extreme weather conditions is what is described as climate variability. This normally takes place on all the spatial or the temporal scales that are normally beyond that of individual weather events. In most cases, the climatic variability does not take place because of the systematic events but rather due to random events. When this takes place, scientists describe it as random variability or in simple terms, noise. With the case of periodic variability, there is relative occurrence and in distinct modes with regards to variability or rather the climatic patterns. There is normally a close relationship or rather a correlation between the climate oscillation on the earth surface and the astronomical factors which include; barycenter changes, solar variation, cosmic ray flux, and the cloud albedo feedback (Shepardson et al., 2011).

The climatic oscillations of the earth are also closely correlated with the modes of heat distribution between the oceanic atmosphere and the climate system. In other instances, the historical, current, and the paleo climatological natural instances tend to be masked by a number of significant world occurrences. Such occurrences include irregularities in the climate proxy data, impact events, positive feedback processes, volcanic eruptions, and the anthropogenic eruptions of GHGs.

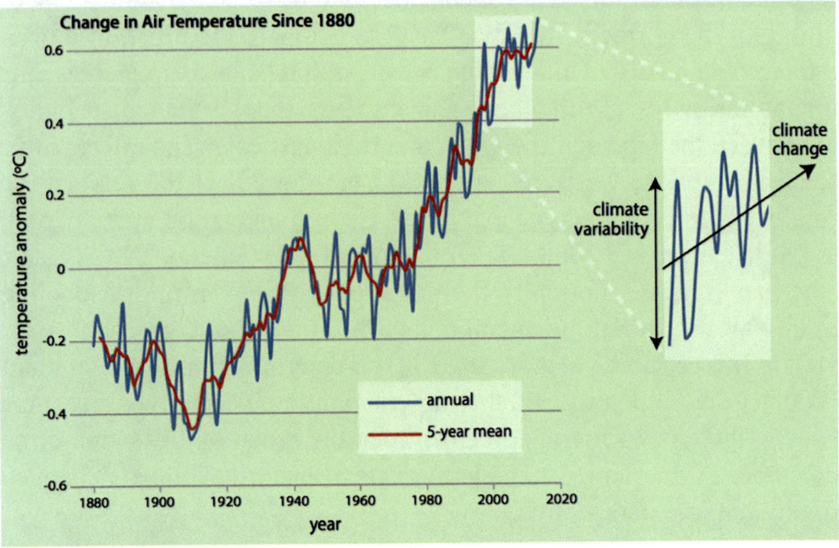

Figure 2.3: Climate variability describes the mean state of the climate and the possibility of extreme weather conditions.

Source: https://scied.ucar.edu/learning-zone/how-climate-works/climate-variability.

Over the past few years, the description of climatic variability and the related term, which is climatic change, have shifted for a while or rather for a number of times. While in the current time, the description of climate change is long term and that which is normally brought about by human activities, in the 1960s, what we understand to be climate variability was what was described as climatic change. In that period, the description was that describing the inconsistencies and the anomalies in the weather patterns.

2.3. CLASSIFICATION OF CLIMATE

Classification of climate is the art of formalizing systems that recognize, clarify, and simplify climatic similarities and differences that exist between geographic regions. This practice enhances the scientific comprehension of the various climates. The classifications majorly depend on vast amounts of environmental collected data that has been sorted and grouped. This enables patterns between interacting climatic processes to uncover. Classical Geek times are among the earliest profound classifiers of climate. These schemes divided the earth into latitudinal zones based on the world's latitudes, i.e., the

Equator, the Tropics of Cancer and Capricorn, and the Arctic and Antarctic circles and on the length of day. In the mid-19th century, the first modern climate classification was published. It included the maps of temperature and precipitation over the Earth's surface, which success led to the development of methods of climate grouping that used the two variables simultaneously.

Several schemes have broadly differentiated types of climatic classification as either empiric or genetic methods. In making use of observed environmental data such as temperature, humidity, precipitation as well as minor quantities derived from them, including evaporation, empirical methods are best applied. On the contrary genetic methods classify climate based on its causal elements, activities involved, and the characteristics of all factors such as circulation systems, fronts, jet streams, solar radiation, and topography, among others. Which give rise to the temporal and spatial climatic data patterns. Two approaches are largely considered while making generic classification, which are based on air mass analysis and based on characteristics of surface energy budget (Shapiro and Park, 2015). Therefore, while empirical climatic classifications are mainly descriptive of the climate, generic methods, on the other hand are explanatory. However, empirical classifications are largely adopted due to their enormous applicability. Other widely accepted climatic classifications schemes include Koppen and Thornthwaite.

In the year 1900, Koppen came up with the most popular empirical climatic classification that has been revised severally thereafter. The scheme used critical values of temperatures of the warmest and coldest months as well as rainfall of the wettest and the driest months. The climatic divisions particularly coincide with vegetation divisions (Zaval et al., 2014). Koppen divided the world climate into five principal groups and other subgroups. The major groups include tropical rainy climate, dry climates, humid mesothermal or warm temperate rainy, humid micro thermal or cold forest climates, and polar climate, which are represented by the letters A, B, C, D, E, respectively.

In the tropical rainy climate, the temperatures do not exceed 18°C on the coolest day. The main sub-group under this classification includes (Af), which is the tropical rain forest or generally the equatorial climate. Throughout, there are warm temperatures with a mean value not exceeding about 27°C coupled with abundant rainfall of about 250 cm annual average. This climate is suitable for luxuriant vegetation and mainly prevails over the Amazon basin, Zaire basin, and South East Asia. The other subgroup is the

Aw, which is tropical savannah climate. It is indicated by the wet summers due to conventional rainfall experienced and dry winter with an annual rainfall of about 160 centimeters. The average mean temperature ranges at 23°C (Figure 2.4).

Figure 2.4: The tropical rainforest climate during the rainy day.

Source: https://study.com/academy/lesson/tropical-rainforest-biome-facts-lesson-for-kids.html.

The vegetation is tropical grassland or savannah with majorly sparsely populated deciduous trees. In this climate, floods and drought are very common cycles (Soutter and Mõttus, 2020). This climate mainly prevails over Sudan, Veld plateau, and the tropical grasslands of Australia. Monsoon type is another subgroup of group A that is represented by the letters Am. This climate has a seasonal reversal of winds that are associated with alternate periods of rainfall and drought with a short dry season. It is mainly experienced over the pacific coast of Colombia, Guinea coast of West Africa, Southeast Africa, south and southeast Asia, and northern Australia. The dry summer is the last but not least climate under the group represented by letters. As it is a hard-to-find climatic type which mainly prevails over some rain shadow areas along the eastern coast of southern India in Tamil Nadu and Orissa that remain dry during summer monsoon and receive winter rainfall or monsoons after retreating.

The second group of climates known as the dry climate widely known for its potential evaporation that exceeds precipitation and constant water deficiency is encountered. The first of its subgroup is the desert or low altitude climate that is represented by the letters (Bwh). Sub-tropical high-pressure region with a mean annual temperature of 38°C are experienced accompanied with scanty and erratic rainfall. The vegetation in this climate varies with the soil type. This is experienced in the southwest USA, North Africa, West Asia, and the Thar Desert, as well as in central Australia. The other subgroup is the mid-latitude deserts represented by (Bwk) letters. They are seen to be very similar to the low latitude desert conditions. These climatic conditions are experienced majorly in Takla Makan in China and Gobi Desert in Mongolia. Semi-arid and Steppe is the last subgroup under this group which is represented by (Bsh) and (Bsk) letters, respectively. The climatic conditions are found in the deep interiors of landmasses such as Eurasia and North America. Their average annual temperatures are around 21°C with rainfall mean at about 30 centimeters. These areas are dry due to an interior location and absence of mountain barriers across the path of prevailing winds (Singh & Purohit, 2014).

The humid mesothermal or the warm temperate rainy group tend to experience mild winters while the mean temperature of the coldest month is below 18°C but above –3°C. However, during the warmest month, the temperatures are above 10°C. One of the subgroups is the humid subtropical which is also termed as Chain type climate that is represented by the letters (Cfa). The climate is hot with humid summer and a mild winter which has an average temperature of about 20°C annually. The rainfall is however well distributed at about 100 centimeters. Commonly, heavy hurricanes and typhoons are experienced. This climate triumph within 25° to 45° latitude on the east coast in both hemispheres, for example, in Southeast USA, Southern Brazil, Uruguay, Argentina, and Southeastern Africa, Eastern coastal belt of Australia, eastern china as well as in Japan. Marine west European climate is the second sub-group under group C that is represented by the letters (Cfb). This subgroup climate is characterized by on shore oceanic influences. Short cool summers and mild winters with average annual temperature around 10°C and rainfall is 140 cm is experienced (Shepardson & Charusombat, 2011). Weather is however variable and unpredictable under these climatic conditions. It prevails around between 45° latitude and 65° latitude on the west coast in the two hemispheres, for example, in the Western Europe, narrow coastal belt in North and South America, south-eastern Australia, and New Zealand. Mediterranean climate is the final subgroup under the group(C) and is represented by (Cs) letters. Mainly the climate is warm and has dry summer with average temperature ranging from 20°C to 27°C attributable to the subtropical high-pressure conditions and mild winter of temperatures ranging from 4°C to 10°C. Its rainfall is from low pressure cyclones which are about 40 centimeters to 60 centimeters annually (Kenis and Mathijs, 2012). This climate is found within 25° and about 45° latitude on west coasts in both hemispheres such as over central California, central Chile, the Mediterranean region, southern South Africa, south-eastern along with south-western Australia.

Humid microthermal, also known as the Cold forest climates, is the group D classification of climate. Here there are severe winters while the temperature of the coldest month is at least below –3°C. The warmest month, however, is above 10°C. Subgroups that exist under this classification include the cool east climate, which is indicated by letters (Df). Humid summer with average temperatures of about 25°C is the order of the day. The temperatures are influenced by tropical maritime air masses where cold winter temperature ranges from a mean of –4°C to 0°C. There is a variable precipitation, which is accompanied by convectional rainfall during summer and snowfall in the winter period. The climate is prevalent on regions between 45° and 65° latitude on east coasts, over north-eastern USA, lower Danube plains, Korea, Japan, and in northern China. The other subgroup is the taiga climate, which is indicated by the letters Ds. The climate has short summer of temperatures between–10°C and about 15°C. Long and cold winters coupled with long precipitations are the order of the day. This is because the climate is influenced by continental polar air. Taiga climate is mainly experienced over the belts from Alaska to Newfoundland and from Norway to Kamchatka peninsula sub-Arctic areas. These classifications of climate refer to the softwood coniferous forest cover. The continental type of climate marks the last classification subgroup according to Koppen climate classification that is represented by the letters (Dw). Temperatures under this classification of climate range from 10°C to 21°C with short and cool summer. There is variable rainfall mostly during the summer periods as well as snowfall during the winter period (Shapiro & Park, 2015). This is attributable to the long and cold winters that have temperatures below 0°C. The climate prevails in deep interiors of the continents around in Taiga and the mid-latitude deserts over Poland and the Baltic states, Russian plains, northern states of USA together with the southern states of Canada.

The last but not least group of classification is the polar climates, which are represented by the letter E. The warmest temperatures under this climate are below 10°C. Additionally, this kind of climate does not experience any warm season. There are only two subgroups including tundra climate, which is the (Et) group. This subgroup is mainly found over coastal fringes of the Arctic Ocean. The climate is characterized by short, cool summer, long, cold winter as well as meager precipitation. Taiga vegetation is limited by this climate. The second subgroup is the Icecap climate indicated by the letters (Ef). The climatic regions under this classification are permanently covered with snow over all seasons. This is attributed to the cold temperatures whereby the warmest month has temperatures below 0°C on average. These conditions are found over the poles and the interiors of Greenland. Lastly is the Highlands climate that is particularly found over the mountainous regions of Rockies, Andes, Alps, and the Himalayas. Vegetation zoning from foothills upwards are similar to those of latitudinal change. The climate is generally high in isolation with low temperature, low pressure, and high precipitation along with larger diurnal ranges. The larger diurnal ranges are experienced tremendously at higher altitudes (Kump, 2011).

The most outstanding aspect of Koppen's classification scheme is that it uses measurable and visible physical attributes of climates such as temperature and precipitation of an area while combining the interactions with the vegetation grown as the basis of the classification. The scheme is practical and very convenient as it uses letter symbols to denote various characteristics. However, it has some limitations as it ignores other factors in the classification, which are essential in climate attributes such as cloudiness, wind, rainfall intensity, currents, and above all, the air masses as they form the basis of modern climatology. Another limitation is that it is difficult to explain the existence of different vegetation types within the same climatic division and similar vegetation types in different climatic divisions in this classification method.

Thornthwaite rational classification improved Koppen classification that was limited by the lack of rational basis for selecting temperature and precipitation values for different climatic zones or areas. Thornthwaite improved this classification by introducing the water balance concept in his classification scheme. He compared the potential evapotranspiration, PET that is defined as the amount of water that could evaporate and transpire from a vegetated landscape without restrictions other than the atmospheric demand with precipitation and computed 'moisture index.' This considers the water surplus indicated by letter(s) in his formula and water deficit indicated by the (d) letter, which occurs in different seasons in most places. Here, water surplus means seasonal addition to subsoil moisture that is being used by the crop at a reduced rate of transpiration during deficit season. Thus, the climate of a place is defined based on either potential evapotranspiration whereby it combines loss of moisture from vegetation surface as evaporation with transpiration or seasonal variation of effective moisture or average annual thermal efficiency. Moreover, Thornthwaite assumed that a surplus of 6 inches of water will counteract a deficit of 10 inch and thus gave higher weightage to the humidity index in his formula of computing the moisture index (Lineman et al., 2015).

According to Table 2.1, Thornthwaite considered moisture indexes to be above zero as there is moist climate in the areas represented by letters (A, B4, B3, B2, B1, and C2). On the other hand, he considered dry climate moisture index to be below zero, which were indicated by the letters (C1, D, and E). According to Nacada, Thornthwaite gave subdivisions to express the levels and extents of dry periods under moist climate and the levels and extent of moist periods under the dry climate. Boundaries were taken in the arithmetic progression of moisture index with a common difference of 20 cm.

Table 2.1: Climate Types and Moisture Index in Thornthwaite Classification

Symbol	Climate Type	Moisture Index	Broad Group
A	Per humid	100 and above	Moist climate
B4	Humid	80 to 100	Moist climate
B3	Humid	60 to 80	Moist climate
B2	Humid	40 to 60	Moist climate
B1	Humid	20 to 40	Moist climate
C2	Moist sub-humid	0 to 20	Moist climate
C1	Dry sub-humid	–20 to 0	Dry climate
D	Semi-arid	–40 to –20	Dry climate
E	Arid	–60 to –40	Dry climate

Source: Nanda (2018).

Thornthwaite considered potential evaporations as thermal efficiency index and used that as a thermal limit in his climatic classification scheme as indicated by Table 2.2. Boundaries were taken in arithmetic sequences progression of the thermal efficiency index with a common difference of 14 cm as expressed in Table 2.2.

Table 2.2: Subdivisions of Moist and Dry Climates in Thornthwaite Classification

Moist Climate	Description	Aridity Index	Dry Climate	Description	Humidity Index
R	Little or no water deficiency	0–10	D	Little or no water surplus	0–16.7
S_1	Moderate summer water deficiency	10–20	S_1	Moderate summer water surplus	16.7–33.3
W_1	Moderate winter efficiency	10–20	W_1	Moderate winter water surplus	16.7–33.3
S_2	Large summer water deficiency	>=20	S_2	Large summer water surplus	>=33.3
W_2	Large winter water deficiency	>=20	W_2	Large winter water surplus	>=33.3

Source: Nanda (2018).

There exist significant similarities between Koppen and Thornthwaite climate classification schemes. Firstly, the two schemes are based on the empirical investigation as they abandon the genetic way of description. In capturing the vegetation aspects of climate, Koppen considered vegetation to be a direct indicator of the totality of climate; while on the other hand, Thornthwaite gave indirect recognition to the vegetation attributes through the concept of evapotranspiration such as the transfer of water from plants to the atmosphere. Besides, both have used temperature and rainfall as basic atmospheric elements controlling climate in any region or area. Although, Koppen considered the absolute values of critical climatic determinants, temperature, and rainfall as recorded at different places while Thornthwaite. On the alternative, considered the values of critical climatic determinants, temperature, and rainfall through thermal efficiency and precipitation effectiveness which he calculated by using evaporation and moisture indices. This scheme has undergone a number of improvements be Thornthwaite after the original in the year 1931, in the year 1933, he modified it further after. Thereafter he made several changes in the method of the year 1933 in the year 1948. Later in the year, 1955 Thornthwaite made other modifications of his 1948-year work to a more improved version, which is currently more widely used in applied climatology studies of various kinds (Mika et al., 2018).

An example of climatic subdivision by both Koppen and Thornthwaite is the climatic subdivision of India. Diverse climatic conditions across India have been attributed to the vast expanse of India across the latitudes and longitudes as well as the presence of large mountain ranges. This is also coupled with long rivers along with the wide range of soil types or kinds. In this region, the monsoon also plays a very significant role in modulating the seasonality of the climate. Placing many considerations on the deep association of natural ecosystem including the agricultural, forestry vegetation, and other human economic activities in addition to livelihood in general with the climate is imperative to characterize the climate at regional scale. This is done by adopting some rational basis for the identification, characterization, and judicious exploration of potential natural resources in the whole ecosystem. The National Commission of Agriculture also adopted this climatic classification of climate in India and the agro climate.

Based on the Koppen climate classification method, India comprises of 8 climatic regions, which are the monsoon type with short dry season represented by the letters (Amw), monsoon type with dry season in summers described by the letters (As). Others include the tropical savannah type represented by (Aw) letters, Semiarid steppe climate assigned to (Bshw) letters, Hot desert type described by (Bwhw) letters, Monsoon type with dry winters represented by (Cwg) letters, Cold humid winter type with short summers indicated by (Dfc) letters and lastly the polar type which is represented by the alphabetic letter (E). The Koppen's scheme, according to the map indicated that the climate represented by (Amw) type of climate prevails over the western coast of India, which is found mainly in the south of Goa. This type of climate is characterized by dry summers is experienced along the Coromandel Coast. The dry climate prevails over the two parts in India. The interior peninsula, Rajasthan, and parts of Haryana have (Bshw) type of climate, while the extreme western Rajasthan experiences climate represented by (Bwhw) letters type of climate characteristics. Most of the peninsular plateau has tropical savannah type of climate, which is represented by the letters (Aw) type of climatic conditions (Milfont, 2012). The plain of India falls under the warm temperate type of climate with dry winters, which is represented by the letters (Cwg). Moreover, the north-eastern India happens to fall under climate conditions represented by the letters (Dfc) type of climate. Here, the winters are generally very cold and humid, while on the other hand, the summers are mostly short. Kashmir and the adjoining mountain ranges are the regions with polar type of climate represented by letter (E) according to Koppen climate classification.

Alternatively, Thornthwaite climatic classification scheme of the regions in India falls under only six different categories of groups. This includes the Per humid represented by the letters (A), Humid represented by the letters (B4 to B1), Moist sub-humid represented by the letters (C2), Dry sub-humid represented by the letters (C1), Semiarid represented by the letters (D) and last but not least the Arid represented by the letters (E) climatic classification. As indicated by his classification, it is depicted that India has both moist and dry climates. Under moist climate, the per humid region represented by the letters (A) lies along the west coast of India, south of Goa and some parts of north-eastern India. The humid climate represented by the letters (B) prevails all along the coast adjoining regions of the west coast, northern part of West Bengal, and the neighboring parts of north-eastern India. The moist sub-humid represented by the letters (C2) climate is found along the Western Ghats and over Orissa and West Bengal areas (Schuldt & Roh, 2014). The Ganga valley and north-eastern parts of Central India have a dry sub-humid type of climate, which is represented by the letters (C1). The peninsular interior, western Madhya Pradesh, Haryana, and Punjab have a semiarid type of climate represented by the letters (D). The regions of Saurashtra, Kuchchh, and Rajasthan in India, on the other hand, experience an arid climate represented by the letters (E) according to Thornthwaite climatic classification method.

Given that agriculture is the most important activity in India as it contributes majorly to her economy, unlike other sectors in the country, agro-climatic zones, it has been necessary to characterize the climate experienced by the various areas in India based on the per crop production requirement in those areas or regions. Agriculture not only contributes significantly to the Indian economy but also is the main source of livelihoods to a majority of citizens, especially those that are located at the rural areas and remote areas. The agro-climatic zone areas would entail classification according to the land units in terms of major climates, particularly suitable for certain crops, vegetation, and cultivars. This activity of characterization enhances planning which is essential at scientific management of regional resources to meet the food, fiber, fodder, and fuelwood without adversely affecting the status of natural resources and environment, according to the Food Agricultural Organization (FAO). The various different agro-climatic zones of India has been accomplished successfully by different institutions such as the Indian Council of Agricultural Research (ICAR), Planning Commission, National Agricultural Research Project (NARP) under ICAR, National Bureau of Soil Survey and Land Use Planning (Agro-ecological regions by the NBSS and LUP). These organizations have achieved this goal through research and based on past studies performed by different scholars (Summerhayes and Zalasiewicz, 2018).

There are four climatic zones in the African continent, which are distributed into Saharan, Sahelian, wet tropical, and equatorial climatic types. Wikipedia reported that the climate in Africa has various ranges such as the equatorial climate, the tropical wet and dry climate, and the tropical monsoon climate. Other climates include semi-arid climate, which is also known as the semi-desert and steppe, the desert climate also known as the hyper-arid and arid as well as the subtropical highland climate, which are based on the combination of temperature, precipitation, and evapotranspiration. Temperate climates are said to be rare across the continent except at very high elevations and long fringes. In fact, the climate of Africa is more variable by rainfall amount than by temperatures, which are consistently high, unlike in other continents. African deserts are the sunniest and the driest parts of the continent, owing to the prevailing presence of the subtropical ridge with subsiding, hot, dry air masses. Africa holds several heat-related records. To start with the continent has the hottest extended region year-round, the areas with the hottest summer climate, the highest sunshine duration among other extremes that owe to Africa's position across equatorial and subtropical latitudes in both the northern and southern hemisphere areas (Schellnhuber, 2008). This allows for several different climate types to be found within its regions. Given that the continent mainly lies within the intertropical zone between the tropic of cancer and that of Capricorn, it has high humidity density levels. Precipitation intensity is always high, making the continent always hot. Warm and hot climates are the order of the day all over Africa, especially in the northern part which is marked with arid and experiences high temperatures. In the whole of Africa, only the northernmost and the southernmost fringes of the continent have a Mediterranean type of climate. Just as the equator runs through the middle of Africa, so does the Tropic of Cancer and that of Capricorn. This attribute makes Africa to stand out as the most tropical continent in the world.

At the equator, the hot and dry climate prevailing in the area makes Africa the most vulnerable continent to climate change. This could also explain the difficulties that most African countries if not all have tried to overcome over the years of climate change. Temperatures are hottest within the Sahara regions of countries such as Algeria and Mali and coolest across the south and at elevation within the topography across the eastern and northwest sections of the continent. The hottest average temperature on

Earth is at Dallol in Ethiopia, which averages a temperature of about 33.9°C experienced throughout the year. The hottest temperature recorded within Africa, which was also the world record, was about 57.8°C at Libya in 1922. Although, this was later proven false, being that the readings were derived from an inaccurate reading of a thermometer. The lowest temperature measured within Africa was –24°C at Morocco in 1935. Nevertheless, the major part of Africa experiences extreme heat during much of the year, especially in the deserts, semi-deserts, steppes, and savannahs. The African deserts are among the said hottest places on Earth, with the likes of the Sahara Desert and the Danakil Desert, located in the Horn of Africa region (Peter, 2018).

In the south west, the low-level easterly African jet stream is considered to play a crucial role in the monsoon of Africa. This has been an aid in forming the tropical waves especially in March across the tropical Atlantic and the eastern part of the Pacific during the warm season. The jet is said to exhibit both bar tropic and baroclinity instability, which produces synoptic scale, westward-propagating disturbances in the jet known as African easterly waves, or tropical waves. A small number of Mesoscale storm systems embedded in these waves develop into tropical cyclones after they move from West Africa into the tropical Atlantic, mainly during the August and September months in the year. When the jet lies to the south of normal during the peak months of the Atlantic hurricane season, tropical cyclone is thereby suppressed in its formation. The great parts of North Africa and southern Africa as well as the whole horn of Africa mainly experience hot desert climate or the hot semi-arid climate for the wetter locations. Annually, the rain belt across the continent moves northward into Sub-Saharan Africa buy the month of August then passes back southward into south-central Africa by March. Many studies in the past have recorded that areas with savannah climate in sub-Saharan Africa, for example in Ghana, Burkina Faso, Darfur, Eritrea, Botswana as well as in Eritrea they all have a distinct rainy. Other reports have indicated that El Nino results in drier than normal conditions in southern Africa from the month of December to that of February and alternatively, wetter than normal conditions in equatorial East Africa over the same season.

Snow is an almost annual occurrence on some of the mountains of South Africa, including those of the Cedar berg and around Ceres in the South-Western Cape. Skiing, including snowboarding in the Cape, is a hit-and-miss affair, both in terms of timing of snowfalls, and whether there is sufficient snow to cover the rocks. Snow is a rare occurrence in Johannes's burg. Besides, snow regularly falls in the Atlas Mountains in the Maghreb as well as in the Mount Kenya and Mount Kilimanjaro in Tanzania in the African continent (Philander, 2008). There have been, however, permanent glaciers on the Ruwenzori Mountains on the border of Uganda and the democratic republic of Congo. However, by the 2010s, the glaciers were in retreat, and they are under threat of disappearing through rising temperatures experienced.

While Africa's climate change is increasingly becoming a serious threat among the humans living in the region, anthropogenic climate change is already a reality in this continent. According to the report on Africa's climate in helping decision-makers to make sense of climate information, it reported that Africa's climate is an interim product of the Future Climate for Africa program, which seeks to identify the gaps in knowledge, and fill those with robust, evidence-based information. The Future Climate for Africa program includes leading researchers and institutions from across Africa, in collaboration with peers and peer institutions in Europe and the UK. The UK Department of International Development (DFID) and the Natural Environment Research Council (NERC) fund the work.

In conclusion, climate classification is an effort to recognize, clarify, and simplify climatic similarities and differences between geographic areas in order to enhance the scientific understanding of climates. The earliest known climatic classifications were those of classical Greek times which divide the earth into latitudinal zones with the most popular empirical classifications of modern times are of Koppen and Thornthwaite. As Koppen designed the system based on his experience as a botanist, his main climate groups are based on what types of vegetation grow in a given climate classification region. In addition to identifying climates, the system can be used to analyze ecosystem conditions and identify the main types of vegetation within climates (Rothe, 2015). Due to its link with the plant life of a given region, the system is useful in predicting future changes in plant life within that region. The Koppen climate classification system has been further modified, within the Thornthwaite climate classification method in the middle 1960s that was revised by 1980. The Thornthwaite climatic system of climate classification as indicated in his Indian climate classification system, sought to create a more refined middle latitude climate zone, which tackles one of the major criticisms of the Koppen system group represented by letter (C) climatic classification (Strefler et al., 2014).

Importantly, all climates are assigned a main group (the first letter). All climates except for those in the *E* group are assigned a seasonal precipitation subgroup (the second letter). For example, (Af) indicates a tropical rainforest climate. The system assigns a temperature subgroup for all groups other than those in the (A) group, indicated by the third letter for climates in *B*, *C*, and *D*, and the second letter for climates in group (E). For example, (*C*fb) indicates oceanic climates, which are prevailed with warm summers. Lastly, as reported by most studies in the past on climate classification, climates are classified based on specific criteria unique to each climate type.

2.4. CLIMATE CHANGE

This is a phenomenon that takes place in the world that are brought about by large-scale shifts in weather patterns. Global warming is an example of climate change and is caused by human emissions of GHGs. The changes in the climatic conditions, however, started even before the dawn of human beings. However, over the years, humans have had the greatest contribution towards the factors that lead to the changes in the climatic conditions. The conditions actually accelerated since the mid of 20th century, which was the age when the industrial revolution took a turn for the better. Even though the industrial revolution has been seen to bring about the improvement of the economy in various parts of the world, their impact on climate is negative (Pleijel, 2009). Ideally, we can say that the change in the climatic condition has been towards the negative side with the temperatures rising and thus the melting of the ice. The melting of the ice is rather a negative situation since there will be a rise in the levels of water bodies in all parts of the world. With the rise in water levels, there will be destruction of some countries, which are a few meters above sea level.

The occurrence of global warming is largely brought about by the GHGs emissions, which is composed of mainly carbon dioxide. The composition of carbon dioxide is over 90% in the greenhouse emissions. The main source of these emissions is the burning of fossil fuels, which include natural gas, coal, and oil for the purpose of acquiring energy to run various machines. Additionally, other sources of GHGs include manufacturing agriculture and deforestation. All the scientific bodies of national and international standings are in agreement that the major cause of climate changes are the human activities. In most cases, climate change is accelerated due to the changes in climate change. This involves either the rise in temperatures or when the temperature has been tampered with (Robinson et al., 2009). This can be caused by the various climatic feedbacks, which include an increase in water vapor, which is also considered to be a GHG, loss of sunlight reflecting snow and ice cover, and the changes to the land and ocean carbon sinks. These increases and the tempering of the temperature has become a hard situation to deal with given the fact that the world wants to evolve with the introduction of new machines each and every day. Such machines require energy in order to run, which involves the burning of fossil fuels.

The increase in temperature, which is almost as twice the increase in the global temperature, brings about the expansion in deserts and further leads to the occurrence of wildfires and common heat waves. The arctic is where the rise in temperature is amplified. This has led to the melting of the permafrost, the loss of the sea ice and has also led to the glacial retreat. Temperature that are warmer than usual bring about an increase in evaporation which leads to weather extremes and intense storms. In a way, these rises in temperatures bring about conventional rainfalls, which are rather more intense, than the normal rainfall. These occurrences bring about changes in the ecosystem, which include the reallocation of most of the living species and the extinction of some species. Taking for instance, the occurrence of wildfires in the natural forests that is home to most animal species in the world that are almost extinct. If such animals did not have the ability to run or rather reallocate on time, they may be caught by the wildfire and burnt to extinction. These changes normally take place in the mountains, the coral reefs and the arctic. The Arctic, for instance, is a region that is highly affected given that most species of animals living in the region are adapted to the cold. The rising temperatures are making the region warmer and warmer and thus the death of the majority of the species of animals. The threat of climate change, however, is not only on animals but humans are also in danger. The threat to humans is because they are at risk of going through after scarcity, flooding, food insecurity, economic losses, displacements, extreme heat, and infectious diseases (Schellnhuber, 2008). As such, the World Health Organizations has brought to the attention of individuals around the world that the changes in the climatic conditions may so far be the greatest health threat in the world. Researches have even gone to the extent of agreeing with the fact that a number of deaths have been brought about by changes in the climatic conditions. The claims are that even though there may be efforts to minimize warmings in the future days, the current effects of warming in the ecosystem will continue for centuries to come. Some of the effects that will take place even for years to come include the problem of ocean acidification, rising ocean temperatures and rising sea levels. However, his is not a discouragement to the organizations that are trying their best to prevent global warming. They are encouraged to continue with the same spirit so as to prevent the world from further decay.

It is with no doubt that the current levels of warming are felt by individuals all around the world. As such, some of the governmental and intergovernmental agencies are coming together to try to find measures that will protect their countries and the citizens. One such agency is the intergovernmental Panel on Climate Change which has brought forth some series of reports that involve the application of certain projects that are quite significant as the impacts of warming continues all over the world and beyond. With an increase in warming, tipping points are at a risk of being triggered, of which the points are considered to be very critical thresholds. In order for individuals to be able to adapt to climatic changes, there will be a need for individuals to adapt to the circumstances and find ways to mitigate the problems brought about by climatic changes. Mitigation involves limiting climate changes and, by extension, limiting their effects on the environment. One of the ways, which is actually, what really needs to be done, involves the reduction of the GHG emissions and further removing them from the atmosphere. Some of the methods known to men to aid in the mitigation of greenhouse emissions include the development and the use of low-carbon energy sources such as solar and wind, reforestation, and sorest preservation, a phase-out of coal, and enhanced energy efficiency. As for adaptation, this involves the ability of individuals around the world to adjust to the actual or rather the expected climatic condition (Revesz et al., 2014). Examples of such measures include better disaster management, the development of more resistance crops, improved coastline protection and assisted colonization. Both the measures are considered to be very important in the process of trying to averse the risk of severe widespread and irreversible impacts of climate change.

There are pacts that have been put in place in order to keep the warming at bay. One of the pacts include the Paris Agreement which the nations had the agreement that the warming need to be kept at minimal levels, that is under 2.0°C. The agreement was that this could only be maintained through mitigation measures. Even though there are a number of pledges that have been made during the enactment of the agreement, the fears are that by the end of the century, the warming will still be at 2.8°C. In order to limit warming to 1.5°C the world needs to have emissions by 2030 and further have the ability to achieve near-zero emissions by the year 205 (Revesz et al., 2014). The advice to individuals all over the world is to understand the mitigation measures and further adapt them. Furthermore, to achieve better results, the world needs some form of discipline. Apart from adopting the mitigation measures, the world needs to improve in terms of their activities. The activities performed by individuals should be those that are good for the ecosystem. Take, for instance, instead of using wood from the forest as a source of fuel; individuals can decide to create their own manmade forest in order to avoid deforestation. The world we live in will be with us for a very long time, and unless individuals are disciplined, they will be the species of living things that will become extinct. Discipline is required, and further introduction of industries and machines should be put aside in order to prevent further destruction of the environment. All animals are important in the ecosystem as each of them play an important role. As such, individuals should work hard to preserve their existence in the atmosphere.

2.5. IMPACT OF CLIMATE CHANGE ON HEALTH

Climate change is among other natural or manmade stressors that have found their way in influencing human health and contributes significantly to infectious diseases in numerous ways. Climate change leads to the emission of other new health threats by intensifying the existing health threats in the human body. This affects those that are very young or old in terms of age, the poor in the society, and those that are located in the regions that undergo climate change more often. The infections are attributed to the disruptions that occur, which may be physical, biological, and though the ecological systems, among other disturbances (Robinson et al., 2009). The major human effects include increased respiratory and cardiovascular diseases, injuries, and premature deaths among infants. This disease is attributable to the extreme weather conditions during climate changes. In the occurrence of climate change, there are changes in the prevalence and geographical distribution of food and water-borne illnesses coupled with other infectious diseases and severe threats to mental health.

The prevent causes of climate change driver of negative impacts to human health include increasingly frequent, intense, and long-lasting extreme heat, which worsens drought, wildfire, and air pollution risks. This increasingly worsens the frequent extreme precipitation, intense storms, and changes in precipitation patterns that lead to drought and ecosystem changes. This changes harm through various ways. Firstly, climate change increases ground-level zone or particulate matter (PM) air pollution in some locations. The increased ground-level zone or air pollution as a result of climate change is associated with many health problems such as diminished lung function, increased hospital admissions and emergency room visits for asthma, and increases in premature deaths, to mention but a few. The factors that are associated with the increased zone formation include heat, concentrations of precursor chemicals, and methane emissions, while PM concentrations are affected by wildfire emissions and air stagnation episodes, among other factors in the environment.

Climate change leads to an increase in the global temperatures. This causes increases in premature deaths related to worsened ozone and particle pollution among the human population. Every year there are approximately 1,000 to 4,300 estimated premature deaths due to regulatory controls caused by the combined ozone and particle health effects. Particularly there is less certainty in the responses of airborne particles to climate change than there is about the response of ozone. Health-related costs of the current effects of ozone air pollution exceeding national standards have been estimated at over $6.5 billion nationwide in the U.S assessment as of 2008 (Pleijel, 2009). Climate change, resulting in more frost-free days and warmer seasonal air temperatures, can contribute to low development and growth in crops that are depended upon by human for food consumption. This is because they shift the flowering time and pollen initiation from allergenic plant species, and increased CO_2 by itself can elevate production of plant-based allergens that consequently affect humans. The high pollen concentration that is enhanced by the longer pollen seasons increases the chances of humans in contacting allergic sensitizations and asthma episodes. In children, these symptoms are mainly reflected by their downscaled school productivity (Rothe, 2015).

The simultaneous exposure to toxic air pollutants consequently worsens allergic responses. Rainfall and temperature extremes add to the decline in air quality, which includes the growth of indoor fungi and molds. This increases asthma disease. In the instance that increased pollen exposures occur, patients and their physicians will face increased challenges in maintaining adequate asthma control. Another effect of climate change is the increasing vulnerability of many forests due to wildfires. Long periods of record high temperatures are associated with droughts that contribute to dry conditions and drive wildfires in some regions. The wildfire smoke contains PM, carbon monoxide, nitrogen oxides (NOx), and various volatile organic compounds (VOC) that are ozone precursors and can significantly reduce air quality to humans living in the areas with these fires. Past studies have shown that smoke exposure increases respiratory and cardiovascular hospitalizations, emergency department visits, and medication dispensations for asthma, bronchitis, chest pain, chronic obstructive pulmonary disease, respiratory infections, and medical visits for lung diseases. Besides, smoke has been associated with hundreds of thousands of deaths annually and especially leading in people living in those areas with fires.

Extreme heat seasons are said to have for a long time threatened public health in the United States. Many citizens have suffered dramatic increases in death rates during heat waves. Deaths were mainly attributed to heatstroke and other related conditions. Cardiovascular disease, respiratory disease, and cerebrovascular disease were also some of the deadly infections (Kenis & Mathijs, 2012). According to many hospital admissions, heatwaves lead in causing cardiovascular, kidney, and respiratory disorders among these patients. Fortunately, some of the risks of heat-related sickness and death have diminished in recent decades, possibly due to better forecasting, heat-health early warning systems, and/or increased access to air conditioning for the U.S population. However, extreme heat events from time to time remain a cause of preventable death to this nation. Urban heat islands, combined with the increased urbanization, are projected to increase the vulnerability of urban populations to heat-related health impacts in the future by past studies.

Additionally, milder winters resulting from a warming climate can reduce illness, injuries, and deaths associated with cold and snow among a population. Increases in both extreme precipitation and total precipitation also have contributed to increases in severe flooding events in certain regions, floods, leads to approximately about 100 deaths per year, while most are mainly due to drowning. The very dangerous floods are associated with tropical storms that are killers of humankind. In addition to the immediate health hazards associated with extreme precipitation events when flooding occurs, appear once a storm event has passed. Elevated waterborne disease outbreaks have been reported in the weeks following heavy rainfall mainly, although other variables apart from floods may affect these associations. Water intrusion into buildings can result in mold contamination that manifests later, leading to indoor air quality problems. Buildings damaged during hurricanes are especially susceptible to water intrusion (Schuldt and Roh, 2014). This may cause to some extent, diarrhea due to contamination of drinking water by the floods that displace the waste from its disposal. Populations living in damp indoor environments experience an increased prevalence of asthma and other upper respiratory tract symptoms, such as coughing and wheezing as well as lower respiratory tract infections.

Lastly, climate change is one of the factors that influence the distribution of diseases borne by vectors (such as fleas, ticks, and mosquitoes, which spread pathogens that cause illness to humans and other insects and animals living in the ecosystem.

2.6. VULNERABLE POPULATIONS AT RISK FROM EFFECTS OF CLIMATE CHANGE

In 2016, the blue-cut wildfire swept through San Bernardino in California and led to the damaging of over 100 homes. The situation was, however, worsened by drought-like conditions that were driven by climatic changes. At this time, the residents in the region struggled after the incident, mostly affecting individuals with fewer resources. These individuals included the senior members of the society who normally have a difficult time adjusting after the occurrence of such incidences. According to the news anchors in the region, the senior members of the society normally do not have the luxury of rebuilding a particular place unless other members of the society assist them. This is because such individuals are vulnerable to various environmental occurrences since they are not able to deal with the smoke or the heat that most people encounter, especially when building a house. The vulnerable members of the society do not however, include only the seniors. Other vulnerable groups include children, people with disabilities, low-income-earning individuals, minorities, and pregnant mothers. As already discussed, the health effects brought about by climate changes in a particular region, which normally relate to extreme weather conditions, are particularly devastating to these groups of individuals. Such individuals are normally more susceptible to health complications, especially when they have pre-existing health conditions or, in other instances, be living in areas that are poorly maintained and do not promote well-being (Schuldt and Roh, 2014).

As already described, the extreme weather conditions have the capability of harming the population of the world in a number of ways. Take, for instance, a rise in the temperatures. They have the ability to prolong the stay of mosquitoes, which are the leading cause of malaria, longer than average allergy season, and individuals are always put at risk of vector-borne disease, which includes the West Nile virus. The most heavily researched section concerning the vulnerable members of the society and how they are affected by climatic changes in extreme heat conditions. With extreme heat conditions, heat waves are normally experienced. When this occurs, and there are cases of air pollution, the individual or rather the particular region becomes vulnerable to heat-related illnesses such as heat stroke, heart attacks, and asthma.

The children and the elderly individuals are the ones that are normally at high risk since it is quite hard to control the body temperatures, unlike when dealing with individuals that fall under the category of the middle-class spectrum. In most instances, the elderly normally has a difficult time as compared to the other individuals. This is so because of their age, and some of the complications that are experienced by these individuals include diabetes, obesity, and respiratory disease and are normally made worse when they are exposed to heat. In most instances, these individuals normally depend on the presence of caregivers to give them care and further ensure that they are safe. One even finds that in most cases when the elderly is in a particular place and are in a long-term care, in case of a power outage in the building, the individuals are normally evacuated in order to protect them from health complications brought about by extreme heat conditions (An et al., 2018).

Some of the climate change-related effects that may lead to individuals in the elderly home to be evacuated includes flooding, which is normally very disadvantageous to the vulnerable members of the society.

2.7. WHO RESPONSE TO CLIMATIC CHANGES?

Many policies and individual choices may play the role of reducing the GHG emissions and provide benefits to our health. Taking for instance, the use of cleaner energy systems and the safe use of the public transportation system to promote active movement. Such active movements include the use of bicycles and walking instead of using private cars. By doing this, individuals reduce carbon emissions from using vehicles and further cut down on the amounts of polluted air that is produced within the household. Air pollution is one type of pollution that causes higher death rates around the world, and by implementing the mitigation measures, their effects on the health of individuals will be reduced. During a sit down in 2015, the WHO brought about a work plan on climatic changes and their effects on the health of the individuals. Some of the components that were included within the work plan included:

1. **Awareness Raising:** This involves the case of taking the responsibility of informing individuals of the dangers of climatic changes. At this stage, the climatologists and the health scientists

should take the initiative to inform the individuals all around the world of the health complications and what they should do in order to protect themselves from such risks. Furthermore, they need to inform members of the public of the opportunities that are known to promote their health and further show them the ways of cutting down carbon emissions.

2. **Partnership:** The WHO has put in place an initiative to coordinate with partner agencies within the system of the United Nations. By doing this, it will be ensured that the topic of climate change and its effects on the population in the world is on the agenda of the day. They should look at such an issue as an important matter that needs to be discussed in the forums conducted by the United Nations (Al-Amin et al., 2013).
3. **Support for the Implementation of the Public Health Response to Climate Change:** This involves assisting the various countries in the world, including the vulnerable ones; these include third-world countries. The agency agreed to work within its power to reduce health vulnerabilities to the changes in climatic conditions while at the same time promoting activities that reduce carbon emissions in the world.
4. **Science and Evidence:** This is important in every aspect as it helps provide the data and the evidence on how dire the effects of climatic changes can be. Furthermore, it helps provide the relationship between health and climate change and further help in the development of agenda in developing global research.

Chapter

3

Health of People

CONTENTS

3.1. INTRODUCTION

The well-being of every living thing is important. For the success of everyday activities in a living organism, be it an insect or a human being, all the components that makeup the living organism ought to be functioning properly. However, just like a coin has two sides, living organisms do face challenges with their well-being; hence, they are forced to seek out ways which will restore their well-being; and also, in order to ensure that the chances of having to face such challenges are minimal, they have to stick to certain practices which maintain their well-being at top-notch levels.

The term health has been used for centuries, the only difference being that, over the centuries, different terms were used to refer to the same thing which health does refer to. However, the definitions of the term health do vary from one location to another; though with further scrutiny of the different definitions of the term health, they all mean one and the same thing (Bassu et al., 2014).

Originally, the term health was used to refer to the biological well-being of living organisms; meaning, all the body organs that make up the entire body system of a living organism are functioning properly. However, with time, people realized that health entailed more than just the biological well-being of living organisms. To human beings, health entailed their emotional well-being, their physical well-being, their social well-being, among others.

To the people, the term health is used to refer to their physical well-being in that all their body parts are functioning properly; their emotional health in that their mental capabilities are okay and they have no mental disabilities; their social well-being in that they can interact normally with fellow human beings; and that they do not have any diseases in them.

3.2. HEALTH OF PEOPLE

As it has been stated earlier, the appropriate meaning of the term health entails, physicality, emotionality, and the sociality of an individual. However, our current generation of the 21st century views health as more of an ability than a necessity. By ability, it means the feeling or the perception and individual has about themselves in reference to being healthy; whereas by necessity, it means the requirement by a doctor for an individual to be mentally and physically fit. With perception of one's health in terms of ability; an individual is able to declare themselves healthy even in the presence of diseases within their body system. Basically, perception of health in terms of ability, focuses more on the ability to feel healthy, which is a factor that promotes resilience towards diseases. This is because an individual is able to fight whichever illness they could be suffering from with a positive attitude of "I will beat this illness;" and by focusing less thoughts on their physical conditions and more thoughts on the positive side of life. The addressing of health in terms of the ability to feel, supports the saying that: as a man thinks, so shall he be.

3.3. FACTORS THAT DETERMINE THE HEALTH OF PEOPLE

With health, there exist several factors which have to be analyzed so as to rule out that an individual is healthy. Some of the factors go hand in hand with the original meaning of the word health, whereas other factors address health in terms of the ability to feel and the view that an individual has towards health (Betts et al., 2011).

All the factors which determine an individual's health status fit into three categories. One of the categories is on the social and economic environment of people. It addresses factors to do with the ability of an individual to interact well and peacefully with other people; their financial status and how it affects them in their day-to-day life. The second category is that of physical environment. With physical environment, the focus is more on the biological well-being of an individual. Some of the environments in which people live in are not healthy in that they might be unclean; which is a factor that affects to the biological well-being of people. The third category is about the characteristics, lifestyle, and behavior of an individual. The ways in which we people behave does contribute a lot to their health status.

The subdivisions of the above-mentioned main factors which determine the health of people are numerous. One of them is the income and social status of an individual, which falls under the category of social and economic environment. An individual who has a low income might find it difficult to provide themselves with the basic needs such as a balanced diet to boost their immune system; warm clothing to protect them from cold climates which affects one's health; money to settle hospital bills which cater for regular checks up which ensure that an individual has a clean bill of health; among other difficulties. As for social status, some people might feel pressured to fit in certain social class. With too much pressure, some end up suffering from ailments caused by lots of stress (Kellstedt et al., 2008).

Social support network is a factor which fits under the category of social environment. Not all individuals who have people that believe in them; people that can support them in all ways possible just to see them prosper in life and achieve their dreams. Lack of social support networks, to basically any individual, is a great tragedy. As the saying goes, "No man is an island." We all need people to be there for us, to enjoy life with, to talk to. With no social support, an individual is exposed to stress-related diseases which, if not treated in advance, could lead to serious damages to any human being.

The factor of education falls under the category of an individual's characteristics. It addresses the issue of the ability to read and write; the ability to interact with individuals of diverse culture for with education, people are exposed to different cultures and able to understand them; the ability to express oneself in the most respectful and appropriate way. Lack of proper educations lowers an individual's self-esteem levels which pushes most individuals to withdraw from the public. With withdrawal, an individual is exposed to different ailments: from stress-related ailments; to ailments caused by lack of self-care such as not feeding on a balanced diet, not keeping the body fit; among other sicknesses (Benjamin et al., 2017).

The working conditions of people do influence their health a lot. For instance, an individual who is under a lot of pressure to deliver quality work within a short period is likely to neglect themselves. An individual who has neglected himself or herself is likely to forget to do the essential things such as feed on a balanced diet, unclutter their minds from negative and unnecessary thoughts, keep themselves fit through exercising, among other practices which boost someone's health. Such neglecting habits lead to slow deterioration of an individual's health.

The factor on social environment do have their role to play when it comes to people's health. Naturally, human beings are less stressed about life when they interact with people who understand them better and make them feel happy and relieved about their day-to-day tensions of life. This is why people tend to prefer certain social environments which help them get away from the busyness of life. However, some individuals tend to be unlucky, this is because, most of the social environments in which they reside-in or work-in do not relieve them of their stress, but instead, they add more tension to them. The lack of peaceful and happy social environments, more so the frequently visited social environments to an individual, do lead to deterioration of one's health. This is because stressful social environments lead to stress-related ailments, among others.

Physical environments can either boost or lower the health of an individual. An individual living in a clean and safe environment will be more protected from diseases. This is because a clean environment will be free of disease-causing microorganisms such as the likes of bacteria and fungi. An unclean physical environment is more prone to the manifestation of disease-causing microorganisms such as the likes of bacteria, fungi, and even sometimes viruses. This is because disease-causing microorganisms will be attracted to the dirt which they usually feed on. Therefore, it is important for people to ensure that their physical environments are sparkling clean to lower their chances of being infected.

The way people carry out their personal health practices and how often they do it determines their health status. This is a factor that falls under the category of people characteristics and behaviors. An individual who sticks to a strict schedule about their personal health practices such brushing teeth, bathing, exercising among other practices; will be healthy compared to an individual who neglects his or her personal health practices. This is because, as the name suggests, personal health products exist so as to boost the health of an individual (Bose, 2010).

The upbringing of an individual does affect their health either in a positive or in a negative way. An individual who was brought by parents who focused so much on health will value the importance of health, and develop behaviors that focus on boosting their health. However, an individual who has been brought in an environment which paid less attention to its health conditions is likely to lead a life of not valuing health practices, which is a habit that exposes them to illnesses. Therefore, it is important for all parents of guardians across the world to develop a culture that boosts the health of people.

The genetic composition of an individual does also determine their health. Some of the sicknesses that exist in this world are passed from parent to child genetically. Other ailments could be as a result of their biological structure. Therefore, biology, and genetics is a health determinant factor that falls under the category of individual's characteristics. Ailments such as autism could be as a result of the biological structure of a child at birth. Ailments such as autism could be controlled, but the scientific world is yet to discover a way of healing autisms. Despite the fact that biological disorders are natural, they still fall under the category of sicknesses which affects an individual's health. The same does apply to genetic ailments, which are passed from one generation to another in a family lineage.

The ways in which people are provided with health care services does affect their health. We live in a world where there are different types of people; some might be rude, impatient, disrespectful, careless, disregarding among other negative traits; whereas others might be polite, respectful, caring, compassionate, loving, among other positive traits. All these people are present in all parts of the world and in all fields. When it comes to the delivery of health care services to individuals, a person might come across a health care worker who will treat them just for the sake of the money they get from the services they provide, basically, a health care worker who does not care. Being treated by such an individual improves the chances of more deterioration of one's health, for the health care worker will not pay much attention to the sick individual, which could lead to wrong diagnosis hence wrong treatment. However, an individual who is lucky enough to be treated by a caring and compassionate health worker will be served well, and the chances of wrong diagnosis which leads to further deterioration of one's health are minimal (Buell, 2014).

Gender is a health determinant factor that is overlooked by many. Both genders are prone to diseases, but the fact is; either genders are more prone to certain illnesses. For instance, to the females who have menstrual cycles each month, the chances of them suffering from anemia are high if they do not feed on meals that add iron to their body. Males are also more prone to certain ailments as compared to female.

The way of life is a factor which determines the health of people. This is because some of the practices practiced in different cultures could either boost health or lower it among people. For instance, the practice of female genital mutilation (FGM) in several culture, does affect the health of the victims. Some of the girls who undergo FGM struggle with recovery hence more deterioration of their health. Such culture ought to be discouraged to ensure that the health of people is boosted. Other cultural practices such as the feeding on raw meat does expose people to different ailments.

3.4. HEALTH ISSUES

People are faced with several issues which affect their well-being. Basically, all the issues that affect health are interlinked with the factors which determine the health of people. This is because, if the factors are dominant, people will be more prone to health issues.

One of the most common health issues is diseases. All the determinants that affect the health of different individuals lead to one thing, disease; with the only difference being in the type of disease they cause. Diseases are categorized into two; one category is the communicable diseases, and the other one is the non-communicable diseases. With communicable diseases, it means that they are easily passed from one individual to another; hence they are highly contagious; whereas non-communicable diseases are diseases that are not passed from one individual to another, meaning they are not contagious.

Communicable diseases are numerous. They include ailments such as Tuberculosis which could be passed from one individual to another through air molecules, or through consumption of non-pasteurized milk; the acquired immunodeficiency syndrome (AIDS/HIV) which is passed from one individual to another through the sharing of body fluids such as saliva, blood, among others; Malaria which is passed from one individual to another by the female anopheles mosquito; among other communicable diseases. Communicable diseases are said to be the cause of millions of annual deaths across the world (Brulle et al., 2012).

Non-communicable diseases are the most common ones despite the fact that they are not contagious. They include the likes of cardiovascular diseases which are diseases related to the heart organ; cancer which affects any part of the body from brain, to bones to blood; diabetes; chronic lung diseases, hypertension caused mainly by stress and consumption of high intakes of salt; among other non-communicable diseases. Non-communicable diseases are said to be the cause of almost 30 million annual deaths across the world.

The other health issues which either leads to death or prolonged suffering of people is malnutrition. As the name suggests, malnutrition is as a result of lack of certain nutrients in the body of an individual. Malnutrition is caused by the feeding of non-balanced diets among people. The practice of feeding on non-balanced diets is mostly caused by the lack of enough funds to purchase food, leading to people feeding on basically any meal they come across so long as they will not feel hungry. Malnutrition is common among children; however, in serious situations, it is ruled out to be present among adults. For decades, malnutrition has been a major cause of high mortality rates among young children.

Body injuries fall under the category of health issues. This is because, once a person is injured, the functioning of the physical organs are affected hence deterioration of their health in one way or the other. Body injuries range from broken bones, to fractures, to burns among others. A broken bone takes time to heal, and in some cases, more so among adults, it might never heal, leaving them with the option of having to use prosthetics. Fractures do also take time to heal, but with them, there are high chances that an individual will go back to being normal despite the age group they are in. Burns do heal, and with time the scars disappear; however, serious burns may take time to heal and the scars rarely disappear. In our current generations, scars caused by serious burns can be eliminated through plastic surgeries which replace the burned skin with an artificial human skin. Most of the body injuries are caused by accidents, some of which could be avoided and others cannot be avoided. Body injuries do reduce the quality of someone's life, more so the body injuries which leave scars or never heal completely; and in some cases, body injuries do lead to death. The lifestyle of an individual is considered to be a health issue. This is because our health is largely reliant on the kind of life we choose to live. Some of the lifestyle's individuals choose include the life of smoking cigarettes, excessive consumption of alcohol, intentional deprivation of sleep, poor dieting either through overeating or overrated intermittent fasting; among other practices (D'Amato and Akdis, 2020). For instance, the lifestyle of frequent smoking slowly deteriorates the health of the lungs leading to fatal ailments such as cancer; excessive consumption of alcohol could contribute to body injuries caused through accidents or lead to ailments such as liver cirrhosis which is a fatal disease; lack of enough sleep affects the proper functioning of different body organs such as the brain due to lack of enough rest; poor dieting could lead to obesity which is a major cause of cardiovascular disease or even malnutrition to the adults due to lack of necessary nutrients in the body. The lack of a balanced lifestyle is a major threat to the health of people.

The lack of sufficient health care system in all parts of the world is a major health issue. This is because, with a limited number of health care facilities, individuals will not be able to be treated whenever they are in need of medical attention. Most people around the world die because they lacked access to health care services either because they could not afford them or because health care facilities were not within their reach. Due to lack of access to health care systems, ailments that could have been treated early in advance and completely being cured end up being chronic for with lack of early treatments they continue worsening, such as the likes of cancer.

3.5. HEALTH AMONG DIFFERENT AGE GROUPS

Health issues affecting people do differ from one age group to another. This is because the immunity of people in different age groups do differ. A child will have a weak immune system, and their lifestyle will be different; however, as the time progresses, and as the child grows from being a toddler to a teenager to a young adult to an adult, the immune system gets stronger and also their lifestyle does change. It could be ruled out that the older and individuals gets, the stronger their immune system; however, the irony is in that, it reaches a certain point in life, where, instead of the body's immunity getting stronger, it starts to weaken, till old age close to death. Such is life.

3.5.1. Newborn Babies

The health of a newborn child is quite delicate. This is because a newborn baby is exposed to a new environment which they had not been used to, and now they have to adapt to the world of breastfeeding, being protected from coldness, having to bathe among other new activities they have to adapt to. Due to this sudden change, some newborns find it difficult to survive in this new world. Most of the newborns who do not manage to pull through their first days after birth are affected by the kind of environment and the care they receive from the word go (Dai, 2011).

When it comes to taking care of newborn babies, it is important that quality carc bc given to the newborn. Necessary equipment should be available in case a newborn is stillborn or delivered prematurely. Mothers expecting to deliver should be able to take good care of themselves to ensure that, at the onset of breastfeeding, the newborn is able to feed on quality breast milk which will boost their immunity hence high chances of their survival. Hospital services should be made available to all mothers and affordable for them to ensure that both the mother and the newborn get appropriate care at the time of delivery.

Newborn babies are at risk of being infected with communicable diseases such as AIDS/HIV; hence it is advisable for expectant mothers to undergo testing as early as possible to prevent the chances of transmitting the disease from mother to child during delivery. This is because newborn babies who get infected with communicable diseases such as the likes of AIDS have low chances of survival since their immune system is not yet strong enough to fight the viruses.

3.5.2. Young Children

Young children are also referred to as infants. The difference between infants and newborn babies is that, newborn babies are children which have been born in the last 10 to 15 days, whereas infants are children which are past 15 days since delivery. Infants have the advantage of they will have managed to successfully adapt to the new world; however, they are not yet out of danger because of their weak immune system. An infant or a young child ranges from 15 days since delivery to 5 years old. (Jang & Hart, 2015)

Infants are prone to diseases such as the likes of malaria if they are exposed to environments infested with mosquitos; respiratory tract infections if they are not protected from coldness; diarrhea; measles if they are not vaccinated in advance, among other communicable and non-communicable diseases. Several vaccines have been invented to protect young children from diseases such as diarrhea, pneumonia among other illnesses. It is advisable that lactating mothers embrace the concept of vaccination for their young ones in order to reduce the mortality rates due to such illnesses. For instance, *Streptococcus pneumoniae* which is the microorganisms that causes pneumonia among young children could be prevented with early vaccination to the young ones; *Haemophilus influenza type b,* which is the bacteria known to cause diarrhea, could also be prevented with early vaccination. Some of the illnesses which affect young ones do not have a vaccine; hence it is advisable for parents to seek treatment as early as possible before the illness has advanced in a child's body (Diffenbaugh and Burke, 2019). Still, under vaccination, almost all nations across the world have incorporated routine childhood immunization programs, which have been made free to the public. These routine immunization programs are aimed at reducing permanent infections to children from illnesses such as polio, whooping cough, among other illnesses, which tend to affect young children for the rest of their lives, and in serious cases, some do lead to death.

To boost the immunity in infants, most lactating mothers are advised to feed their infants purely on breast milk for a minimum of six months. This is because breast milk is believed to contain all the necessary dietary requirements for the survival of an infant and for their growth. Past six months, a lactating mother is allowed to feed their child on other diets such as water and soft meals for the continued growth of the child. Breastfeeding does lower the chances of a child ending up to be underweight. To boost the quality of breast milk, most lactating mothers are advised to feed on nutritious meals. If possible, women should feed on nutritious meals from the first day they realize they are pregnant to ensure that the fetus gets enough nutrients while in the womb; and towards delivery, expectant mothers should increase their intake of nutritious meals to ensure that by the time of delivery, the breast milk to be produced is quite nutritious. Nutritious breast milk prevents growth retardation among newborns and infants. Upon introduction of other meals other than breast milk, it is advisable that a lactating mother be more mindful of the diet to ensure that the child does feed on a balanced diet. This is because a balanced diet does protect the child from malnutrition; and as stated earlier, malnutrition tends to be more serious in young children and it could lead to death at a faster rate compared to other ailments.

Young children should be brought up in a safe environment that does not expose them to diseases. For instance, a young child should be able to sleep under the treated net to protect them from mosquitos which cause malaria. In homesteads which rely on charcoal as their source of fuel, should be careful not to expose the young children to smoke for it contains carbon dioxide, which could affect their respiratory system. Young children should be protected from other sources of fuels such as the likes of petroleum products for it is easy for them to consume them assuming them to be water which ends up affecting the health of their digestive systems. All products which recommend to be kept out of reach among young children should be kept far away from children; because, in one or the other, if the child manages to get hold of them, they will play with them and in one way or the other the child's health will be affected from their use (Donnelly et al., 2017).

In most communities, their culture supports FGM among young girls around the age of 5 years. Most of the mortality cases among the young girls are as a result of FGM. FGM does cause acute bacterial infection which could be fatal among young girls; hemorrhage which is excessive bleeding leading to the death of many innocent girls for their immune system is not yet strong enough to persevere excessive loss of blood; and, most of the girls who manage to survive past FGM are at a greater risk of contracting AIDS/HIV. AIDS could also be transmitted to the young girls during the FGM process if the same objects are used for several girls in the group. Therefore, it advisable that such culture be completely discouraged across the globe to save the life of our girls.

The world we live in is slowly turning out to be an evil one. This is because rape cases are even being reported whereby most victims are young children under the age of 5 years. It is advisable that parents be more mindful of who they live their young children with. Rape at an early age does affect the health of the child, from their physical health to their mental health.

Young children do play a lot, for it is believed that "too much work without play makes them dull children." During their play sessions, there is no guarantee that they will remain safe. Most of the times while playing young children end up hurting themselves. As discussed earlier on body injuries, some of the injuries end up healing whereas others change a child's healthy life. So, it is advisable that, during playtime, parents and guardians should be keen on the kind of plays their young children engage in and what are the possible risks from the play of such games.

As stated earlier, the health issue of availability of quality health care services does also affect young children. Most of the medical personal, more so the nurses, are not that well skilled in handling young children and infants; hence, it is advisable that the skills of health care workers be improved from time to time to ensure that the young children receive quality services increasing their chances of recovery (Fountain, 2019).

The issue of illiteracy among the members of a certain community does affect the health of young children. For instance, when it comes to the routine immunization programs, most illiterate parents believe the vaccines to be demonic or even dangerous to their children; and they end up not immunizing their children or opting for traditional methods to treat their young ones. Some of the traditional methods do work, but only in communities where they have gifted herbalist; whereas in other communities, the traditional methods lead to catastrophic effects. Therefore, it is advisable for governments to hold frequent workshops to educate the public on the importance of the vaccines, how they work, and if they have any side effects. This way, the public, both the educated and the illiterate, will embrace vaccines and reduce the mortality rates among the young children due to lack of vaccination. Apart from the conducting of public workshops, governments should invest in surveillance programs. We all know that not all human beings agree with what is being taught to them. The surveillance programs will ensure that the importance of vaccination is emphasized to the parents and guardians who ignored what was taught to them while in the workshops.

3.5.3. Young People

Young people are believed to be individuals between the ages of 7 to 24 years. The category of young people can further be divided into two. The first category being between the ages of 7 to 19 years, which is the category in which young people shift from childhood into the adolescent stage, mostly known to many as the teenage years (Haines et al., 2007). The second category is of people between the ages of 15 to 24 years, which is referred to as the youth. The youth are quite mature and able to make some reasonable decisions about their life. Young people are affected by all sorts of health issues, but the most prevalent ones are under the category of social and economic environments. This is because young people get the chance to be exposed to the outside world away from their parents. It is at this age where most of them are send to boarding schools; from primary boarding schools to secondary boarding schools and later to university boarding schools. With this exposure, young people are exposed to emotional life struggles, for they have to learn to make decisions away from their parents and be willing to embrace the consequences of their decisions; have to learn how to deal with conflicts for at such ages they will have to interact with diverse people who have beliefs and opinions different from theirs; among other exposures which affect their health either directly or indirectly.

Some of the major health issues affecting young people are communicable diseases such as AIDS/HIV, unwanted pregnancies and child marriages; limited access to family planning services; injuries due to either violence or conflicts, among other health issues. As for the unwanted pregnancies, young people are being initiated to sex quite early, more so in our current generation as compared to the previous generations. With this exposure, they are at high risk of ending up pregnant, and due to the fear of rejection and stigmatization; most of them opt for abortion which could lead to death or permanent infertility (Herndon, 2018).

With initiation to sex at an earlier age, young people are exposed to the risk of contracting communicable diseases such as the sexually transmitted diseases, which include the likes of AIDS/HIV, syphilis, gonorrhea, among others. Some of the sexually transmitted diseases such as HIV/AIDS do not have a cure, so the young person will have to cope by use of ARVs which are known to suppress the effects of the HIV virus. However, if the dosage is not followed to the latter, the rate at which the immune system weakens is quite fast and it could lead to them. The contracting of sexually transmitted diseases might discourage young people from seeking treatment early in advance due to the fear of being judged or mocked for engaging in premarital sex; which leads to serious health effects to them due to the untreated disease.

With our current generation, it is high time the society embraces that young people have been initiated to sex or if not initiated, they have been exposed to what sex is. With this understanding, the society will be able to address sexually related issues early in advance to avoid the transmission of sexually transmitted diseases, early teenage pregnancies, among other effects. Young children ought to be taught on sexually related matters for them to fully understand the advantages and disadvantages of engaging in sexual activities. If possible, family planning services should be provided to the youth, who are around the ages of 19 years to 24 years; for at this age, the youth are quite mature to make decisions for themselves and are bold enough to accept the consequences of engaging in certain activities (Hiltner, 2009)

As stated earlier, young people are in the category where they are exposed to all sorts of people; and chances are, they might engage in conflicts due to their diversities, and in most cases, the conflicts could lead to violence. To address the issue of conflicts, young people should be taught from the word go, as early as at the age of ten, on how to address conflicts in a peaceful and avoid violence (Herndon and Whiteside, 2019). Parents should also do their best to ensure that they do not expose their young ones to violent activities, for a child's mind replicates what they see happening at home.

At the ages of young people, most of them develop their beauty characteristics, and they end up being prey to sexual offenders. Young boys and girls end up being prey to rapists. The effects of rape are quite catastrophic to both genders; as for the girls, they are exposed to unwanted pregnancies, sexually transmitted diseases and trauma which ends up affecting their mental health; as the boys, they could be damaged from behind and exposed to trauma which does damage their mental health. To curb the effects of rape, serious actions ought to be taken against sexual offenders. Every government should ensure that they formulate or implement the existing laws such they discourage rape activities.

Child marriages are also a great danger to the health of young people. This is because, with early marriage, a child is initiate to sex and since they in a marriage set up, they are expected to conceive and fill the home with young ones. With early delivery, the young girls become vulnerable to reproductive health complications; which in most cases could lead to permanent infertility or even death. For the young girls who are in marriages, access to family planning services is limited, which increases the risk of health issues; due to continued delivery year after year.

The issue of child marriage should be addressed through the formulation of laws that discourage early marriages. Culprits should be made to face serious consequences for breaking the law and willingly exposing young girls to the serious health issues that come with early marriages. The remote communities which are yet to be civilized should be taught about the importance of civilization and encouraged to do away with their negative cultural practices such as early marriages.

Young people are also initiated to drug abuse. As the name suggests, the abuse of drugs does have negative effects to the health of young people. For instance, excessive consumption of alcohol among the youth does lead to liver cirrhosis. The continued smoking of cigarettes exposes one to lung cancer which is a fatal disease. The use of Miraa increases infertility levels among the male. Use of other drugs such as the likes of bhang and cocaine could lead to mental health issues; more so if they are consumed excessively. Not that the young people do not know about the effects of drug abuse, for they are taught at school as early as at the age of 10 years; hence, the issue of drug abuse among the youth cannot be addressed through workshops. Governments should set up corrective facilities and make them affordable to the public for the rehabilitation of young people. Laws should be formulated and the existing ones on drug abuse implemented to ensure that young people are disciplined and that they serve as an example to the rest; hence discouraging the use of drugs among the youth (Hiltner, 2009).

Friendly health care services only for the youth people should be set up. At least this way, young people will feel confident enough to visit health care facilities for treatment of several ailments such as the sexually transmitted diseases.

3.5.4. Adult Women

Adult women are categorized to range from the ages of 20 to 49 years. During these ages, scientifically, women are believed to have reached their maturity level and able to conceive. In terms of health, adult women are affected by both communicable and non-communicable diseases. The communicable diseases include the likes of HIV/AIDS, malaria, tuberculosis, human papilloma virus (HPV), among others. As for the non-communicable diseases, women are affected by maternal health complications, cancer, anemia, breast cancer, among other illnesses.

As for the communicable diseases such as the likes of HIV/AIDS, adult women are known to be at the stage of marriage. Hence they are sexually active. By being sexually active, they are vulnerable to sexually transmitted diseases such as the likes of HIV/AIDS and HPV. HPV is first transmitted as a sexually related disease and if not treated early in advance, it leads to the development of cervical cancer among women. Cervical cancer has been ruled out to be the major cause of mortality among adult women. However, the good thing is, there is a vaccine which has been invented to help in preventing the contracting of HPV disease. The HPV vaccine does not mean that women have to stop going for regular screening; for regular screening ensures that there no signs of cervical cancer.

Women should be more mindful of ways of preventing themselves from suffering from malaria, more so expectant mothers. This is because malaria can be quite dangerous to expectant mothers. Therefore, they should practice sleeping under treated nets and ensure that the environment in which they reside in is clean; ensuring that there are no breeding grounds for mosquitos (Hiller, 2011). In case of the occurrence of any symptoms related to Malaria, expectant mothers are advised to ensure that they seek medical attention as early as possible to avoid the worsening of malaria.

All women should be vaccinated against tuberculosis; whether expectant or not expectant. This is because tuberculosis has been ruled out to be one of the most fatal illnesses. Even after vaccination, adult women are advised to ensure that they always drink pasteurized milk and that they avoid crowded places. In case of the occurrence of symptoms related to tuberculosis, adult women are advised to immediate medical attention for tuberculosis is curable if treated in its early stages.

As for the non-communicable diseases such as the likes of breast cancer, adult women are advised to go for regular screening to rule out the chances of suffering from breast cancer. With regular screening, any breast cancer cases are able to be treated early in advance ensuring that it does not advance to other stages which tend to be difficult to treat.

Anemia is another common non-communicable disease among women. This is because adult women between the ages of around 15 years to 49 years do have a menstrual cycle; and each month, they will be indeed of plenty of iron nutrients so as to replenish the lost blood. Most women neglect the importance of feeding on meals rich in iron nutrients. With such neglecting, women end up suffering from anemia due to lack of enough blood within their system. The issue of anemia could be addressed through frequent emphasis to women on the importance of feeding on meals rich in iron nutrients to replace the lost blood for iron nutrients help in formatting of more blood in the body.

When it comes to maternal health issues, women should be availed with safe family planning techniques, even in the remote areas. The availability of safe family planning techniques will ensure that women are in control of their birth rates. Increased birth rates among women does weaken their body which ends up weakening their immune system making them more susceptible to other illnesses or even lead to death. With reduced birth rates, and enough time for a woman to recover from one delivery, the rates of maternal deaths among adult women will be reduced (Jang and Hart, 2015). The other advantage of family planning is that the newborn child will have enough time to bond with the mother and develop well for the mother is not under body pressure of providing for a second unborn child while taking care of an infant.

With pregnancy, a woman exposes herself to various life-threatening issues. One, there are chances that the expectant mother could experience a miscarriage; two, during delivery, there is no guarantee that the mother will survive due to excessive labor pain; three, the effects which lactating mothers undergo post-delivery are sometimes life-threatening; among other effects. It is therefore advisable that expectant mothers be taken care of in the most appropriate way. For instance, during the expectancy period, expectant mothers should be able to feed on highly nutritious meals to ensure that they are strong and healthy enough to carry on with the pregnancy. During delivery, the health care workers should attend to women in labor in a delicate way and see to it that they have a safe delivery where both the mother and the new born child are guaranteed of survival post-delivery. During the periods of lactation, lactating mothers should also feed on highly nutritious meals to ensure that they recover from delivery, and are healthy enough to take care of two people (themselves and the newborn). Governments also have a role to play by ensuring that health care services are made available at an affordable cost to the expectant mothers.

Adult women are exposed to violence. This is because most of the adult women are at the age of marriage, and not all marriages are that peaceful. Even if they are peaceful, they are between human beings, and naturally, human beings disagree on several occasions. Due to the conflicts which are present in all marriages, some of the adult women could be exposed to violence from their partners. To address the issues of gender-based violence in marriages, governments should formulate laws which ensure that the culprits to such violent activities face the law (Jacobson, 2012).

3.5.5. Adult Males

Adult males' range between the ages of 16 years to 50 years. In most communities, men are a sign of strength and boldness in life. This means that men are not taken care of as women. Most of the issues to do with men are neglected for the society expects them to be bold enough to fight their own issues and overcome the challenges they face without help. However, due to stress issues, men end up being affected by diseases more than women; yet scientifically, women are the ones expected to be more affected by diseases compared to men. Adult men are affected by both communicable and non-communicable diseases. They affected by HIV/AIDS, tuberculosis, malaria, high blood pressure, poor diet, conflicts, body injuries, drug abuse, and depressive disorders.

As for the communicable diseases such as HIV/AIDS, adult men are in the age of marriage; hence they are exposed to sexually transmitted diseases. Tuberculosis is common among men for they engage more in activities which involve overcrowded places meaning that they will be at a higher risk of contracting tuberculosis. Malaria is a communicable disease common to all genders for mosquitos do not choose who to infect malaria. As for HIV/AIDS, adult males should be mindful of the people they engage in sexual activities with; and for the already infected ones, they should be guided through counseling on how to accept their status and learn to live with by adhering to the ARV dosage. Men should embrace practices that lower the transmission rates of malaria, such as sleeping under treated nets among others.

High blood pressure is common among men, for they do undergo a lot of pressure. From work-related struggles to family-related struggles; for as stated earlier, men are expected to be the heads and bold enough to address any challenge that comes across them and also be able to provide (Kgabi et al., 2014). The issue of high blood pressure among men could be reduced by guiding men to take on responsibilities which they can manage to handle; and embrace the act of seeking for help whenever they are in need of help.

Most men live a life of living for the sake of others instead of living a life for themselves. Programs should be set up to help men in loving themselves and help them understand that they need to care for themselves as much as they need to take care of other people. With such programs, men will be encouraged to develop the habits of feeding on healthy diets, which boost their immune system and make them less susceptible to illnesses.

Due to the societal pressure exerted on adult men, most of them find themselves opting for drugs as a way of reducing stress. The consumption of drugs does have diverse health effects to men such as: suffering from liver cirrhosis due to excessive consumption of alcohol; lung cancer due to the smoking of cigarettes and weeds; infertility in men due to consumption of

Miraa; among other negative effects. Furthermore, the use of drugs makes men to be violent, which could lead to issues such as body injuries, and even murder (Jacobson, 2012). Therefore, it is advisable that programs be setup to ensure that there is pressure balance in the society. With pressure balance, the expectations of adult men in the society will be reduced, and they will be able to live a normal life just like any other human being. The society should be taught that times are changing, and that family or any other leadership responsibility is not solely on adult men. Adult men should also be guided on embracing the habit of opening up and seeking for help when they are in need of help.

3.5.6. Older Men and Women

Older men and women range between the ages of 50 years and above. As it was stated earlier, from childhood to a certain age, approximately around 50 years, the immunity in human beings gets stronger with each passing year; however, as they approach the ages of 60 years and above, their immunity systems continue to weaken with each passing year. At old age, men and women end up being more susceptible to various illnesses.

The factor of aging pushes for the need of high availability of health services among the elderly. They should be taken care of in the most appropriate ways. Their diets should be highly nutritious to ensure that their weakening immune system is strengthened. The environments in which they leave in should be clean to ensure that they are not exposed to disease-causing microorganisms; peaceful to ensure that they do not have to struggle with stress issues; and friendly ensuring that they are not lonely, they at least have people to talk to and spent quality time with. Old men and women should be support in all ways possible, ensuring that they do not lack anything (Kellstedt et al., 2008).

With the best care being given to the aging individuals; their health does boost despite their weakening immune system, and they become less and less susceptible to old age illnesses.

In conclusion, the health of the People is so important. However, just like any other sector, public health is faced by several challenges. One of the challenges is to do with the lifestyle people choose to live; two is to do with social and economic environments of people; and three is to do with the healthcare-associated infections. To overcome such challenges, individuals are advised to be mindful of the lifestyle they choose; try to balance their social and economic environments in which they live in; and do their best to take care of their bodies by feeding on healthy and nutritious meals and taking care of their mental health.

The world is made up of people, and it exists for the sake of people; basically, the world needs people for it to survive, so do people need the world to survive. With such knowledge, people should treasure the world by taking care of the environment they live in and taking care of themselves too.

Chapter

4

Types of Air Pollution

CONTENTS

4.1. INTRODUCTION

At rates above the natural capacity of the environment for dissipation or dilution or absorption, air pollution, discharge of different gasses, subdivided solids, or finely distributed liquid aerosols into the air. The concentration of these particulates in the environment may lead to unwanted health, economic, and esthetic impact. Clean, dry air mainly comprises of nitrogen and oxygen—78% and 21% by volume. Other gases (0.9%) combined with hydrogen, methane, carbon dioxide, methane, helium, and more (exceedingly small), are a combination of 1% (Kenis and Mathijs, 2012). Water vapor is also an ordinary element of the environment although quite variable, normally between 0.01 and 4% by volume; air humidity levels can be as high as 5%, if very humid. Air pollution is an important risk factor for several diseases linked to pollution, such as lung cancer, stroke, COPD, heart disease, and respiratory infections. There are far-reaching human health effects of poor air quality, but they primarily affect the air system and the cardiovascular system. Individual air pollutant responses dependent on the characteristics of the pollutant subjected to, the level of exposure as well as the health and genetic condition of the individual. The Blacksmith Institute World's Worst Polluted Places document 2008 lists indoor pollution and poor inner-city air quality as two of the world's worst harmful pollution problems. The result is 2.1 to 4.21 million deaths per year due to air pollution alone in the outside. Overall, the nation's biggest environmental risk factor is caused by air pollution and the deaths of around 7 million people every year. The cost to the world economy of $5 trillion per annum is approximated to be loss of productivity and deteriorated air pollution life quality. There are diverse applications and strategies for reducing air pollution available. Some of these pollutants are discussed in further sections (Figure 4.1).

Figure 4.1: The mixture between solid particles and air gases in the atmosphere is air pollution. Air pollution: As particles, it may be possible to suspend emis-

sions, chemicals from factories, dust, pollen, and molten spores. Ozone, the gas, is a major contributor to cities' air pollution. It is also known as smog when ozone forms air pollution. Some contaminants in the air are toxic.

Source: https://www.news-medical.net/amp/news/20191128/Air-pollution-could-be-associated-with-more-health-problems-than-previously-thought.aspx.

4.2. CARBON MONOXIDE

Carbon monoxide is one of the six significant air pollutants in the United States and several other countries across the world, which is harmful, colorless, and scentless. When carbon-based fuels, like oil, wood, and coal, consume carbon monoxide in an unfinished or inefficient way. The gas is distributed in the lower atmosphere by wind and atmospheric circulation (called the troposphere). The sources and quantities of atmospheric carbon monoxide shift in various parts of the world and in various seasons. The seasonal changes of carbon monoxide, for instance, in Africa are linked to the widespread agricultural combustion in seasons that move south and north of the equator (Kump, 2011). In other areas of the southern hemisphere, as in South East Asia and the Amazon, fires are a significant source of carbon monoxide pollution.

Because of vehicle and industrial emissions, in East China, Europe, and the United States, the largest proportion of carbon monoxide arise in urban areas. Fires in Russia and North America in large areas may be an excellent component in some years. MOPITT findings often demonstrate that pollution from one continent can be transported through oceans to greatly influence air quality on other continents. Carbon monoxide, as carbon dioxide and methane do, is a trace gas in the atmosphere, and has no direct impacts on world temperature. Carbon monoxide, even so, plays an important role in air chemistry and affects the air's ability to purify itself from many other polluting gasses. It also contributes to the creation of lower atmospheric ("bad") ozone and urban smog in connection with other pollutants and sunshine.

The high CO composition of the respiratory air lessens the quantity of oxygen transportable into critical organs such as the brain and heart in the bloodstream. CO can lead to death, unawareness, confusion, and blurred vision at extremely high levels, which can be done indoors or in other enclosed environments. Outdoors is unlikely to produce extremely high CO levels. Even so, if CO levels are high outside, people with certain types of heart disease can be particularly concerned. These individuals are already less able to get oxygenated blood into their heart when the heart requires more oxygen than standard (Zhang, 2017). They are particularly susceptible to CO's effects during exercise or stress. This can lead to lower oxygen to the heart accompanied with a chest pain, also called Angina, in short-term exposure to increased carbon dioxide.

4.3. OZONE

Ozone is a gas that contains three oxygen atoms (O_3). Ozone tends to happen in the top atmosphere of Earth as well as in the ground. Based on where it is encountered, ozone may be bad or good. Good ozone naturally takes place in the upper atmosphere, termed stratospheric ozone, where it acts as a protective layer that protects us from the detrimental rays of the sun. This advantageous ozone, which sometimes is termed the "ozone holes," was partially damaged with man-made chemicals. The good news is that this hole decreases. Ozone is the major component in "smog" at ground level as it has a harmful air pollutant since it affects people and the environment.

Ground-level or ground-level ozone is not directly released into the air but caused by chemical processes between volatile organic compounds (VOC) and nitrogen oxides (NOx). This happens if chemical, refinery, industrial boilers, power stations, cars, and other sources of pollutants contain chemicals in the presence of sunshine. Ozone is most probably unhealthy during hot and sunny days in urban settings, but in colder months still can reach significant concentrations. Long distances of ozone can also be carried by wind, and high ozone levels can also occur in rural areas (Hiller, 2011) Ozone can damage our health in the air we breathe. Older adults, children, people with asthma and people working outside, external workers, are the most at risk for breathing air which is ozone-binding. Furthermore, people with certain genetic properties and people who have reduced nutrient intake, like vitamins E and C, are more at risk from ozone exposure. High levels of ozone breathing can lead to various answers including bruising, coughing, irritation in the throat, and inflammation of the airways. Lung capability and lung tissue damage can also be reduced. Ozone can make asthma, emphysema, and bronchitis worse, resulting in improved medical treatment.

Higher ozone exposures can influence vulnerable vegetation and ecosystems, such as wildlife areas, parks, and forests. Especially in the growing season ozone can harm vulnerable vegetation. The weather forecasts for manual devices, electronic or printed or television often give rise to air quality forecasts. Everybody could take initiatives to decrease air pollution and to maintain air cleaner and safe measures to safeguard one's health. The clean air act defines ozone as one of the six major air pollutants. EPA calls such "air pollutants criteria," even though their outdoor air levels must be limited according to medical criteria (Figure 4.2).

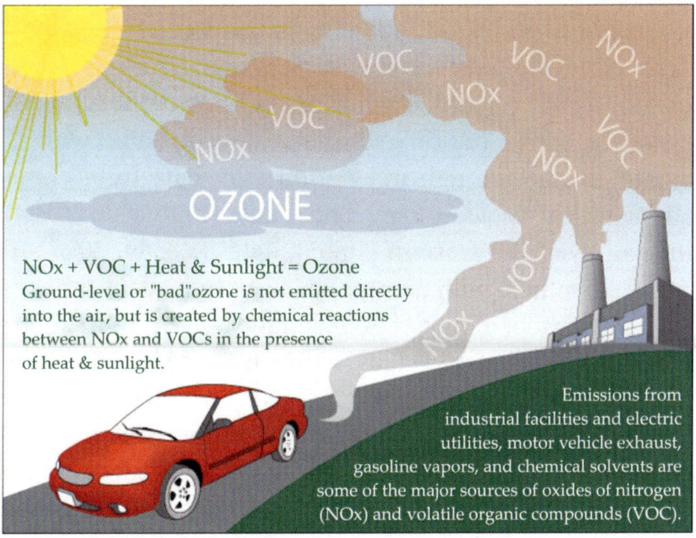

Figure 4.2: Based on the area in the atmosphere, ozone can be "good" or "bad" for health and the environment. "Good" stratosphere ozone helps protect living creatures from the sun's ultraviolet radiation. The subject of this blog is "bad" because ozone levels at ground level can lead to a range of health issues, especially for children, senior citizens, and people of all ages with lung disorders such as asthma.

Source: https://www.epa.gov/ground-level-ozone-pollution/ground-level-ozone -basi cs.

Ozone alone is an atmospheric greenhouse gas (GHG). As a result, rising terrestrial ozone makes a significant contribution to climate change. On the other hand, the warming environment promotes the formation and aggregation of ozone mainly via two physicochemical mechanisms in the atmosphere. Initially, warming weather changes moisture and wind conditions in specific parts of the world, leading to declines in surface cyclones. The resulting stagnation of the atmosphere reduces NOx and

VOC distribution and extends the time to generate ozone reactions (Zaval et al., 2014). Second, the ozone-forming reactions are generally reinforced by higher air temperatures. Based on this climate change in air stabilization (air stagnation) and temperature, the number of ozone-standard days in the Eastern United States alone could rise by 68% by 2050.

A further study anticipated that regional and globally increased ozone changes would enhance soil ozone in most of the US. More specifically, the 95th percentile is projected to rise from 79 ppb by 2012 to 87 ppb by 2050 for the daily maximal ozone of 8 hours. Likewise, an additional predictive analysis has estimated a rise of 0.43 ppb (95% CI: 0.14–0.75) in the average ozone concentration over the 2040s compared with 2000 alone due to climate change through integration with data from climate model output and historical meteorologic and ozone findings across 19 urban communities all over southeastern United States (Zhang, 2017). The ozone season may also continue with climate change. High quantities, of ozone for instance, normally happen in the US in summer. Even so, in 2000 and 2010 in the southeast of the United States, ozone managed to reach the summer level in several Octobers. This was credited to the increased emissions in the warming and drying conditions of biogenic isoprene (VOC ozone precursor) from water-stressed plants. In the long-term drying and warming events could lead, in the areas with substantial biogenic VOC emissions, to an extension of the ozone season, from summer to fall.

4.4. PARTICULATE MATTER (PM)

PM stands for a combination of liquid droplets and solid particles present in the air (also known as particulate pollution). There are large or obscure particles like smoke, soot, dirt, and dust, which can be seen nakedly. Others are so small that only an electron microscope can be detected. Particle pollution includes:

- **PM_{10}:** Particles that are inhaled, with diameters of 10 micrometers and less.
- **$PM_{2.5}$:** Fine particles that are inhaled, with diameters of 2.5 micrometers and less.

These particulates can be made up of hundreds of various substances in many sizes and forms. Certain sources are issued directly, such as building sites, gravel driveways, fields, smocks, or fires. Most particles form in the atmosphere due to the complex chemical processes like sulfur dioxide (SO_2) and oxides of nitrogen that pollute power plants, manufacturing companies, and cars. Microscopic liquid or solid droplets are contained in particulate matter (PM) so small that they are ingested and lead to complications. Some particles with a diameter of under 10 micrometers may enter your lungs deep and some may even enter your bloodstream (Lineman et al., 2015). Of these, the most common health risk is particulates with a diameter of less than 2.5 micrometers, also called fine particles or $PM_{2.5}$. Fine particles, which include many of our preserved national parks and wilderness areas, are also the major causal for limited visibility (haze) in areas of the United States. Inhalable particles are regulated by EPA. EPA does not regulate parts of sand and large dust that are greater than 10 micrometers. The regional and national rules of EPA for reducing PM emissions will enhance the national air quality standards of the Agency for the local and state governments.

A variety of sources which are either natural or linked to human action emit certain particles straight in the air. Marine spray, pollen, dust storms, and bush fires, are natural sources. Human activities include emissions from motor vehicles, industrial processes like wood heaters, roads unpaved, electricity production, incinerators, and stone crushing. These include human activities. Exposure to PM has been linked with several health issues like respiratory diseases (for example, bronchitis and asthma) and cardiovascular disorders. Moreover, the chemical compounds of certain particles were shown to lead to cancer, especially combustion products. These effects for vulnerable populations like young and elderly are often more prominent (Figure 4.3).

Figure 4.3: Subsequent sources release particulate gases. Sub-source examples are power plants and coal fires. Several other prevalent pollution sources may be both primary and secondary; for example, plants, cars, trucks, and buildings.

Source: https://www.cdc.gov/air/particulate_matter.html.

Governments have taken action to overcome the quantity of produced particles and to improve our comprehension of particles' nature and impacts. The following steps are:

- Integrate national standards on fuel quality and promote alternative fuel;
- Support tighter emission regulations in vehicles;
- Development of national diesel environmental measures to improve diesel vehicles' in-service efficiency;
- Working with companies and the territories and the wood thermal industries to improve technology in wood thermal systems, wood thermal operation and conformity to the Australian standard for the emissions of wood thermal PM;
- Working with the state government and territories to ensure that the same size particularly is measured in a true federal monitoring network at all points.

4.5. LEAD

Lead (Pb) is an elementary heavy metal found in both the environment and in products produced naturally. Like suspended particles, lead can be directly discharged into the air. Motor vehicles and industrial sources have been historically the main sources of lead air emissions. The emission of motor vehicles has been reduced through the phasing out of led gasoline; however, lead continues to be used for piston motor aircraft in general aviation petrol. Lead that is released into the air, mainly through contact with polluted soils or other surfaces, can be breathed or consumed (Mika et al., 2018).

The main sources of lead today include lead-acid battery manufacturing and recyclers, utilities, waste incinerators, and lead smelters. Other sources of lead emissions are inclusive of cement manufacturers, glass manufacturers, institutional, commercial, and industrial boilers, copper smelters, steel, and iron foundries, and metal processing. Humans may be subjected directly to air pollution, by inhaling or by incidental plumage, which has faded into the soil or dust from the air. The major site of human exposure to lead initially emitted into the air is the inhalation of lead that is placed on surfaces. Once the body is carried, lead collects in the body and builds up in the bones. Leads can adversely influence the reproductive and developmental systems, immune system, kidney function, nervous system, and the cardiovascular system, depending on the level of exposure. The oxygen-carrying capability of the blood is also influenced by lead exposure (Hiltner, 2009)

Neurological effects in children and cardiovascular impacts (e.g., heart disease and high blood pressure) in adults are the leading effects most frequently available in current populations. Young children and infants are particularly vulnerable to low levels of leadership, which can lead to behavioral problems, shortfalls in learning, and lowered IQ. Lead is chronic and builds up through air-source accumulation, direct waste stream release to bodies of sediment and soil, mining, and erosion in the environment. Blei-point ecosystems show a wide spectrum of adverse effects, such as biodiversity loss, changes in the composition of the community, reduced plant-and-animal growth and reproductive rates and neurological damage in vertebrates.

Lead emissions sources vary between areas. Nationally, metals and ore production, piston-engine aircraft with led air-fuel are the major sources of lead in the air. Other alternatives include utilities, waste incinerators, and manufacturers of batteries with lead-acid (Yeager et al., 2011). The highest air levels of plumage are normally found close to plumbing. During the period 1980–2014, the level of lead in air reduced by 98% because of EPA regulatory efforts, including expelling of lead from the car's gasoline.

After it is brought into the body, lead disburses in the blood and accumulates in the bones throughout the body. The plumage can negatively affect the nervous system, the renal function, the immune system, the reproductive and developmental systems, and the cardiovascular system depending on the level of exposure. The exposure to lead also affects the capacity to carry oxygen in the blood. Neurological effects in children are the most frequent effects in present populations. Infants and young children are particularly sensitive to lead exposure levels that can lead to psychological problems, learning shortfalls and reduced IQ levels (Zaval et al., 2014).

Lead is environmentally persistent and may be introduced to sediments and soils by deposits from sources of air contamination with lead. Certain sources of ecosystems involve direct waste streams to mining and water systems. High environmental lead can result in reduced plant and animal growth and reproductive effects in the vertebrates.

4.6. SULFUR DIOXIDE (SO_2)

The national SO_2 environmental quality standards for EPA are intended to protect the whole set of sulfur oxides from exposure (SOx). SO_2 is the main issue of the larger group of sulfur gaseous oxides and is utilized as a measure (SOx). At atmospheric concentrations much decreased than SO_2 are found to contain other gaseous SOx (such as SO_3). Generally, SO_2 control measures can be predicted to reduce human exposure to all gaseous SOx. The significant co-benefit of the reduction of sulfur particulate formation, like fine sulfate particles can be achieved here (Milfont, 2012). Emissions which generally lead to high levels of SO_2 are also causing other SOx to develop. Fossil fuel combustion at power stations and other industrial installations are the biggest sources of emissions of SO_2 (Figure 4.4).

Figure 4.4: Air sulfur dioxide is 99% derived from human sources. It is an acidic gas which acts in two ways as a pollutant. First, direct to the lungs and airways of humans in which they are irritating and can cause wheezing, density of the chest and pulmonary disease. For individuals who have asthma and some lung diseases, this can be a special problem. The environmental issue of acid rain is also caused by sulfur dioxide (together with various nitrogen oxides).

Source: https://pmgbiology.com/tag/sulfur-dioxide/.

Fossil fuels are burnt by power plants and other manufacturing sites as the biggest component of SO_2 in the atmosphere. Smaller sources of SO_2 emissions are including manufacturing processes like ore extraction, natural sources such as volcanoes, and locomotives, vessels, and other vehicles, and heavy-duty fuel-burning equipment that contain high levels of sulfur. Both health and the environment can be affected by SO_2.

Short-term SO_2 exposure can damage and make it difficult for people to breathe. Those with asthma are sensitive to these SO_2 effects, particularly in children. SO_2 emissions leading to high SO_2 concentrations in the air also generate sulfur oxides (SOx). SOx may react to small particles with other compounds in the atmosphere. These particles contribute to the pollution of particulates (PM). Small pathogens can enter the lungs deeply and can cause health problems in sufficient quantity. Gas SOx can harm trees and plants at high concentrations through the harmful effects of foil and reduced growth. The acid rain may contribute to the damage of sensitive ecosystems by SO_2 and other oxides of sulfur. In other parts of the US, such as our treasured natural forests, and wildlife areas, SO_2 and other sulfur oxides may be able to react with other substances in the atmosphere to form fine particles that diminish visibility (haze). Particle deposits can also stain and damage stones and other materials such as statues or monuments, including cultural objects (Peter, 2018).

SO_2 emissions and SOx-forming pollutants (Saxes') will help states and local governments meet the national air quality requirement for the Agency by national and regional EPA regulations. EPA recognizes areas that do not comply with the EPA SO_2 standards for air quality. The tribal, local, and state governments are developing plans to reduce the volume of SO_2 in the air for these regions.

4.7. NITROGEN OXIDE

NOx is the standard expression in biogeochemistry for the NOx that are most important for air pollution, including nitric oxide (NO) and nitrogen dioxide (NO_2). Smog and acid rain are caused by these emissions, which can influence tropospheric ozone.

NOx gases are most often created as a by-product through the reaction between nitrogen and oxygen as the combustion of fuels occurs. An example of these combustion processes that burn off products such as hydrocarbons, in ambient air, particularly at higher temperatures, can be seen occurring in everyday automobile engines. The NOx released in high-traffic areas, such as in bigger towns and cities, maybe a major contributor to air pollution and lower standards of air quality. A lightning strike also produces NOx gases inherently and naturally.

The expression NOx is chemically definitive of molecules that contain one nitrogen atom and one and sometimes more oxygen atoms. While nitrous oxide (N_2O) is a more relatively inert nitrogen oxide, it has quite several uses such as an oxidizer for rockets engines and car engines. It is also used as an anesthetic in the medical field as well as a propellant for aerosol sprays like whipped cream. Though there is strong evidence suggesting that it could have a direct effect on the ozone layer and is a powerful GHG, nitrous oxide plays a minor role in air pollution.

NOy, when studying the chemistry of inorganic nitrogen compounds, is made up of NOx and NOx compounds. This could include many NOx like nitric acid, nitrous acid (HONO), dinitrogen pentoxide (N_2O_5), peroxyacetyl nitrate (PAN), alkyl nitrates ($RONO_2$), peroxyalkyl nitrates ($ROONO_2$), the nitrate radical (NO_3), and peroxynitric acid (HNO_4).

Nitrogen dioxide is an irritant gas that induces inflammation of the airways and other respiratory problems, especially at large doses.

NO is formed when nitrogen is emitted during the combustion of fuel and then mixes with oxygen atoms. It is denoted chemically as (NO). Nitrogen dioxide is formed when NO then reacts again with another oxygen atom. This is further expressed as (NO_2) in chemistry (Philander, 2008). At normal atmospheric concentrations, NO is not considered a health threat or having any adverse effect however, nitrogen dioxide is. Nitrogen dioxide and NO are both alluded to as NOx as stated above (Figure 4.5).

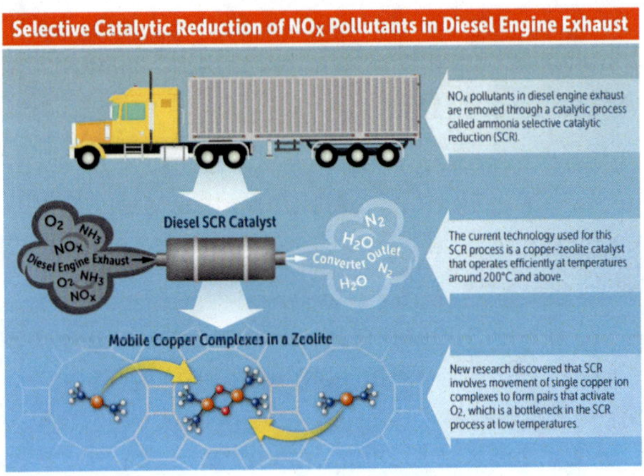

Figure 4.5: Researchers found a new chemical reactivity that would serve to further decrease smog-causing nitrogen oxides in diesel exhalation by improving catalyst designs for contaminant control systems.

Source: https://phys.org/news/2017-08-discovery-catalyst-nitrogen-oxides-diesel.amp.

Nitrogen Oxide gases are involved in the production of smog and acid rain, as well as fine PM and ozone production along the Earth's surface, all of which have harmful health consequences.

The source or production of NOx is more easily apparent than it may seem. NOx is emitted during the process of combustion when nitrogen and oxygen molecules interact in the ambient air, particularly at increasingly high temperatures. The concentration of NOx released into the environment as air pollution can be severe in areas of heavy automobile traffic, such as major cities. Anywhere combustion takes place in the presence of nitrogen, NOx gases are produced, e.g., in all combustion engines; they are also produced naturally by lightning strikes.

From 1970 to 1990, the emission of nitrogen oxide gasses from electricity production remained relatively stable and well monitored. The growing use of natural gas in electric energy production in the early 1990s was a viable solution and replaced coal and oil. Until 2000, NOx emissions were significantly reduced due to cleaner fuel and more efficient power plants (Pleijel, 2009).

Some experiments show that mild NOx concentrations may cause both positive and negative ecological effects, with synergistic relationships with SO_2 being particularly significant. In the context of comparable SO_2 concentrations, there is clear evidence that NO_2 results are more likely to be negative in its effects. Fortunately, SO_2 and NOx ratios have been on the decline in the areas of Europe where these studies were conducted.

NOx can cause a range of significant effect along the respiratory tract, inducing airway inflammation at elevated levels. Long-term exposure can reduce lung capacity, raise the risk of respiratory problems, and exacerbate the effects of allergic reactions. NOx also plays a role in the production of PM and ground-level ozone, all of which are potentially related to negative health conditions.

The impact of nitrogen dioxide on ecosystems cannot be understated. NOx levels above a certain threshold can affect vegetation quite adversely, causing leaf damage and slowed growth. It increases the susceptibility of plants to infections, diseases, and weakens plants making them prone to frost damage. According to an analysis of the effect of nitrogen dioxide and ammonia (NH_3) done in Epping Forest, Essex, pollution by NOx is likely to have a major impact on the forest's ecological functions and general health.

One significant impact of NOx could be on insect populations; research shows that insect pests' function better on plants grown in moderate NO_2

and SO_2 concentrations. The study discovered that local traffic pollution plays a significant role in the overall contribution to the region's nitrous oxide loads, sometimes pushing them to exceeding the critical levels. The critical level for vegetation protection is 30 g/m^3 on an annualized basis (Revesz et al., 2014).

NOx also can interact with other pollutants in the area, especially in the presence of sunlight to form ozone which can damage vegetation at high concentrations.

Note that when discussing the critical level, it is about the ambient concentration of a pollutant in each environment at which adverse direct effects on an ecosystem or species can be observed. The critical load, on the other hand, is the amount of pollutant accumulation to the point which adverse indirect effects on an ecosystem or organisms can be seen.

4.8. CHLOROFLUOROCARBONS (CFCS) AND HYDROFLUOROCARBONS (HFCS)

The protective coating of Earth's ozone, which shields the Earth from sun-generated harmful UV-B rays, is destroyed by halons, chlorofluorocarbons (CFC), or hydrochlorofluorocarbons (HCFCs). The lower Earth environment, changing global climate, is warmed by CFCs and HCFCs. In the warming up of the planet, also hydrofluorocarbons (HFCs) act. The MPCA works with industry, residents, and the governments to decrease the ozone layer and global climate damage caused using HFCs, HCFCs, CFCs, and other related chemicals (Figure 4.6).

Figure 4.6: Ozone is an anti-UV shield around our planet to protect us from high energy rays, which can, when left unimpeded, penetrate the atmosphere, and cause more skin cancer and cataracts, resulting in further degradation, which is harmful to plants and animals.

Source: https://theboar.org/2019/06/cfc-gasses/.

Man-made substances (CFCs, HCFCs, etc.), destroy ozone in the upper atmosphere owing to the use of fluorocarbons and halons (stratosphere). The stratospheric ozone layer enables life by protecting the Earth from sun-generated harmful UV rays. The lower stratospheric ozone concentration permits the surface of the Earth to reach higher levels of UV-B. The loss of stratospheric ozone can lead to possible health and environmental damage, including:

- Enhanced cataract and skin incidence;
- Damage to the immune system;
- Harm to plant life on land and aquatic land;
- Enhanced ground-level ozone formation (smog).

When chlorine or bromine reacts with ozone, the majority of stratosphere ozone depletion is caused. Man-made sources (84%) make up most of the chlorine entering the stratosphere. CFCs and HCFCs are made from natural sources, such as the sea or the volcanoes with 16% leftover. About half of bromine from human-created sources, mostly Halons, enter the stratosphere.

CFCs and HCFCs also trap heat in the lower atmosphere, causing warmth on the Earth and climate and changing weather while acting to destroy ozone. In the lower Earth's atmosphere, HFCs originally appointed to take over from CFCs and HCFCs also absorb and trap infrared radiation or heat. HFCFs, CFCs, and HFCs are a subset of a wider group of GHGs. Taking GHGs together, by the end of the century, the planet is expected to heat by 2,5 to 8°F (Robinson et al., 2009).

The current climate and climate change effects of GHG are estimated at 11.5% by HFCs, CFCs, and HFCs. Some global climate change effects include:

- The level of rise in the sea;
- Extinction and loss of habitat of local natural species;
- Heavy precipitation and flooding are more common;
- Summer's high heat stress;
- The risk of insect and waterborne diseases increases health risks.

As of 1 January 1996, the development or import of CFCs in the United States was prohibited. Equipment placed before 1996 is limited to the use of CFCs. HCFC-22 and HCFC-142b are prohibited from producing or importing for use in new units or applications in the US as of 1 January 2010, even though they will be permitted to produce and import for use in existing equipment until 2019. As of January 1, 2004, it was prohibited to produce or import HCFC-141b for any purpose (Singh & Purohit, 2014.

The majority of HFC uses are phased out under the USEPA rules for new units or applications over an overall schedule starting in 2016 and extending through 2024. Under these Rules, the majority of HFC use will be wiped out between 1 January 2016 and 1 January 2020 in polyurethane and other foams and in retail food refrigerated cases. The use of HFCs in mobiles will conclude with the 2020 model, while bans on the use of HFCs will take full effect on the 1st of January 2018, 2021, 2023, and 2024 for new fires suppression systems, cold storage, residential cooling and building chillers.

EPA regulations do not affect the use of HFCs in existing equipment. Minnesota has stated that roughly 12 million tons of CFCs, HCFCs, and HFCs are present in domestic appliances and car air conditioners. There are rules preventing the release into the environment of CFCs, HCFCs, and HFCs (Rothe, 2015). To obtain a certification technician, proper cooling or recycling materials and records, servants, and devices for appliance and motor vehicle air conditioning units are needed, in addition to the various manufacturing ban.

4.9. ACROLEIN

Acrolein is the most basic unsaturated aldehyde which belongs to a group of organic carbonyl compounds. It is a liquid without color with an acrid, penetrating smell. The scent of burnt fat is generated by glycerol in the burning fat breaking down into acrolein (like how when cooking oil is heated up well past smoke point). It is primarily used more as a biocide and a key ingredient for other chemicals, such as the amino acid methionine, and is manufactured industrially from propylene (Figure 4.7).

Jöns Jacob Berzelius, a Swedish chemist, was the one to first describe and define acrolein as an aldehyde in 1839. At the time, he was working with it as a glycerol thermal degradation component, a substance used during soap manufacturing.

The name, etymologically, is a combination of the words 'acrid' (which refers to the putrid odor) and 'oleum' (contracting to its oil-like viscosity). Acrolein became an essential intermediate product in the industrial development and manufacturing of acrylic acid and acrylic materials later in the twentieth century (Schuldt and Roh, 2014).

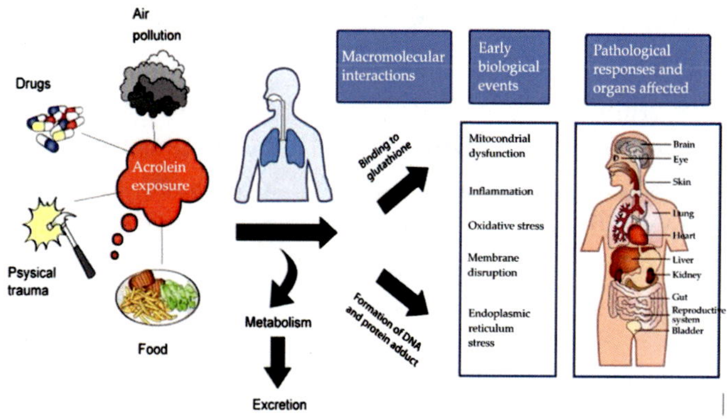

Figure 4.7: Acrolein is a chemical that is employed in the chemical industry as a transitional reactive aldehyde. It is used to synthesize many organic compounds, the formation of methionine and methyl chloride. Acrolein is exposed to smoking, second-hand smoke, wood, and plastic smoke exposure. Firefighters and people in high motor traffic areas may be exposed to higher levels of acrolein through smoke inhalation or automotive exhaust.

Source: https://www.semanticscholar.org/paper/Experimental-studies-on-acute-health-effects-of-and-Dwivedi/8101e65d334ab55126016d38f7f5ac013c1f8d94.

Industrially, acrolein is produced by oxidizing propene. Air is used as an oxygen source, and metal oxides are used as heterogeneous catalysts. This process produces around 500,000 tons of acrolein per year in North America, Europe, and Japan. Furthermore, the transient formation of acrolein is the main source of all acrylic acid. The primary difficulty is combating over oxidation of this acid. Propane is a strong but difficult feedstock to produce acrolein as well as acrylic acid.

When glycerol or glycerin is heated well past 280 °C, it breaks down into acrolein. When glycerol is created as a material in the synthesis of biofuels from vegetable oils or animal fats, this approach is appealing. Glycerol dehydration has been shown, but it has not proved to be effective with the petrochemical path.

Acrolein is generally used as a surface herbicide in drainage canals to combat underwater and floating vegetation, as well as algae. It is used in drainage and re-circulating waters at a concentration of 10 parts per million. In essence, it is used as a biocide in frocking waters and as a scrounger for hydrogen sulfide and other organic sulfides in the oil and gas industry.

Several useful compounds are made from acrolein, having used it as a chemical precursor. When it is around oxygen acrolein polymerizes. Also, when it is at concentrations greater than 22% in water, acrolein polymerizes. The hue and consistency of the polymer are affected by the environment. It can polymerize with itself over time, becoming a transparent, yellow solid. It would turn into a rigid, brittle plastic when exposed to water.

Acrolein is toxic and is a powerful irritant around the skin, eyes, and respiratory passages. Acrolein suppresses the immune system and can encourage regulatory cells, preventing allergy production on the one hand but raising the risk of cancer on the other. One of the chemicals implicated in the industrial contamination of the Kim Kim River in 2019 has been identified as acrolein.

The results of a study conducted by the US Environmental Protection Agency provide strong evidence that human exposure to acrolein is mainly attributed to indoor air with minimal quantities exposed from ambient outdoor air (Schellnhuber, 2008). The research strongly correlated acrolein exposure with temperature and cooking events, and homes with frequent, regular cooking activity.

There are correlations between acrolein gas in tobacco cigarette smoke and the possibility of lung cancer. Acrolein tops the list in "non-carcinogenic health quotient" for tobacco smoke ingredients, accounting for 40 times more than the following compound, hydrogen cyanide. The amount of acrolein in cigarette smoke varies depending on the type of cigarette and the amount of glycerin added. Notably, while filters can decrease the concentration of constituents in conventional smoke, they have little impact on the formation of side-stream smoke, which typically contains acrolein and is breathed by passive smokers.

4.10. ASBESTOS

Asbestos is a fibrous silicate material that occurs naturally. There are six different forms, all of which are made up of long, thin fibrous crystals with several microscopic "fibrils" that can be emitted into the environment by abrasive wear and other methods. Asbestos is a strong electrical insulator and absorber, which has a high heat resistance, so it was used as a construction material for a long time. Asbestos is now widely recognized as a health and safety threat, and its use as a construction material is prohibited in many countries. Asbestos fibers can cause several severe lung problems, including asbestosis and cancer, when inhaled.

While archeologists have discovered evidence of asbestos being used to reinforce ceramic pots as far back as the Stone Age, large-scale industrial mining only started at the end of the 19th century, when producers and contractors began to use asbestos for its advantageous chemical composition.

Asbestos was commonly used in the twentieth century until the 1970s, when widespread awareness of the health risks of asbestos dust led to its ban in most nations across the world for use in conventional building and fireproofing (Strefler et al., 2014). Despite this, at least 100,000 people are believed to die each year from diseases attributed to asbestos contamination, due to the fact that the symptoms of exposure can sometimes take decades to manifest (Figure 4.8).

Figure 4.8: Most people feel it has been completely prohibited by the number of legal actions and the general upheaval over asbestos. However, there are different regulating levels around the globe for this mineral. The US continues to allow asbestos in products to be contained in 1% or less while Australia has completely banned it.

Source: https://www.conserve-energy-future.com/environmental-dangers-asbestos.php.

Even considering the seriousness of asbestos-related infections, the substance has been commonly used around the world, and asbestos is believed to be found in most buildings built before the 1980s. Many developed countries continue to promote asbestos use as a construction material, and asbestos mining continues, with top producer Russia expected to manufacture 790,000 tons in 2020.

When asbestos is ground up, instead of becoming dust fragments, it breaks into millions of fine fibers during the mining processing stage. When asbestos fibers are breathed in, they can cause severe health issues. Since there is no such thing as a healthy level of asbestos contamination, the higher the chance of contracting an asbestos-related illness. The time between exposure and the onset of symptoms can be if 30 years (Summerhayes and Zalasiewicz, 2018).

Asbestosis, a disease caused by asbestos, is a scarring of the lungs that causes respiratory trouble and heart problems. Asbestosis is a disease that affects workers who make or use asbestos materials as well as anyone who is exposed to elevated levels of asbestos. Asbestos inhalation can lead to lung cancer and mesothelioma, a rare cancer of the chest and belly linings. Tobacco smoke has a synergistic effect with exposure to asbestos, increasing the risk of lung cancer. Asbestos has also been linked to stomach, intestine, and rectum cancers.

4.11. CARBON DISULFIDE

The colorless, volatile liquid carbon disulfide is expressed chemically as CS2. The compound is widely used as an essential compound in industrial and organic chemistry as well as an agricultural and chemical non-polar solvent. The technical substance is a yellowish liquid with an unpleasant odor.

Carbon disulfide is an organic industrial chemical used to produce viscose rayon fibers in large amounts. Viscose is a quasi-rayon fabric made from natural wood pulp ordinarily used as a silk alternative because of its almost identical drape and smooth feel. The word "viscose" refers to the wood pulp solvent that is used to create the cloth. Around 20–30 g of carbon disulfide and 4–6 g of hydrogen sulfide is released for every kilogram of viscose developed in this technical process. Coal gasification plants also emit additional carbon disulfide, carbonyl sulfide, and hydrogen sulfide; data on cumulative emissions from these plants is not clear (Shepardson & Charusombat, 2011).

Carbon disulfide contact is largely limited to those employed in the viscose industry's technical processes. The public, however, could be vulnerable to carbon disulfide emissions if they are regularly near viscose plants. Light could decompose some of the carbon disulfide in the air.

Carbonyl sulfide, SO_2, and carbon monoxide are formed through oxidation. Carbonyl sulfide emits a foul odor.

In terms of routes of exposure, inhalation is the main channel of carbon disulfide absorption into the body; however, the chemical can also be absorbed through drinking water and food if it reaches the waterways, especially from the wastewaters of viscose rayon plants.

Carbon disulfide has been related to several health effects of both acute and chronic cases of poisoning. The average threshold limit value (defined as the level to which a worker can be exposed day after day for a working lifetime without adverse effects) is 30 mg/m^3, or 10 ppm. Tingling or numbness, lack of appetite, blurred vision, cramps, muscle fatigue, nausea, neurophysiological deficiency, priapism, erectile dysfunction, hysteria, keratitis, and death by respiratory failure are all potential symptoms. Occupational exposure to carbon disulfide is also linked with cardiovascular disease, particularly stroke.

4.12. BENZENE

Benzene is a colorless or light-yellow liquid at room temperature. It is easily flammable and has a pleasant scent. Benzene evaporates rapidly into the atmosphere. Since the vapor is denser than air, it can settle in low-lying areas. Benzene just partially dissolves in water and floats on top of it (Soutter and Mõttus, 2020).

Benzene is produced from both natural processes and human occupation. Volcanoes and forest fires are two common naturally occurring causes of benzene. Benzene is present in crude oil, diesel, and tobacco smoke as a natural ingredient as well. In the United States, benzene is commonly employed in a wide range of industries. Benzene is used in the manufacturing of paints, plastic, resins, nylon, and synthetic fabrics in some industries. Benzene is also used to produce lubricants, rubbers, dyes, detergents, medicines, and pesticides, among other things. In terms of volume of production, it is among the top 20 compounds (Figure 4.9).

Figure 4.9: Several problems can result in benzene in the air. In the disease control and prevention centers, for example, long-term exposure to elevated benzene levels can produce leukemia.

Source: https://www.carlsonattorneys.com/news-and-update/chemical-storage-fire/amp.

Tobacco smoke, gas stations, motor vehicle fumes, and factory pollution all contribute to existing levels of benzene in the outdoor air. Benzene levels in indoor air are typically higher than in outdoor air. Indoor air absorbs benzene from benzene-containing items such as glues, paints, fabric wax, and detergents.

Benzene levels can be higher in the air at toxic waste facilities or petrol stations than in other regions. Benzene spills from underground storage sites or benzene-contaminated toxic wastes may potentially contaminate well water and other water sources. People who work in businesses that produce or use benzene are likely to be exposed to the higher possible levels. Tobacco smoke is a significant cause of benzene toxicity.

Benzene acts by triggering cells to malfunction. It may, for example, cause the bone marrow to stop producing enough red blood cells, resulting in anemia. It may also affect the immune responses by altering antibody levels in the blood and inducing the depletion of white blood cells. The severity of benzene poisoning is determined by the level, path, and duration of exposure, as well as the affected person's age and pre-existing health issue (Shapiro and Park, 2015).

Within minutes to several hours after inhaling high levels of benzene, people can experience the following signs and symptoms:

- Drowsiness;
- Feeling dizzy;
- Heartbeats that are fast or erratic;
- Aches and pains;
- Tremors;
- Confusion;
- Unconsciousness;
- Death is imminent at very high levels.

Benzene can cause tissue damage and inflammation when it meets the eyes, skin, or lungs, and the above symptoms does not necessarily mean that a person has been exposed to benzene.

Long-term exposure to benzene has a significant impact on the blood. (A year or more of exposure is considered long-term.) Benzene is toxic to the bone marrow and can cause a drop of red blood cells, resulting in anemia. It can also cause heavy bleeding and influence the immune system, making infection more likely (Shapiro & Park, 2015).

In a hospital setting, benzene poisoning is treated with supportive medical treatment. There is no clear antidote for benzene poisoning. The most crucial thing for victims is to get emergency help as rapidly as feasible.

The air exchange between interior and exterior means that indoor levels can be as high as external ones. For indoor air exposure. The indoor air may in some cases be more intense. Benzene levels, for instance, can be considerably higher than in households with attached garage fuel products than those that do not store fuel products. Other instances include houses which use a heating kerosene heater or a cooking gas oven. Tobacco smoke is perhaps the main source of indoor exposure. The average benzene levels in non-smoking homes are about 2.2 parts per milliard, while smokers' homes are 3.3 parts per billion averaged. The concentrations of benzene in bars for smoking range from 8 to 13 parts per billion. Briefly, smoking people consume 10 times more benzene a day than non-smokers.

Chapter

5

Human and Environmental Health Impacts of Criteria Air Pollutants

CONTENTS

5.1. INTRODUCTION

Criteria air pollutants are those that have a specific regulation used when dealing with them. These air pollutants are used in evaluating the level of air quality. These regulations have been placed depending on the impacts of these pollutants in the environment. With criteria pollutants, the standards are set to ensure good air quality is maintained. Most of the criteria pollutants affect the ozone layer. Most of these pollutants are gases. Standards controlling the use these gases are developed by the EPA. They have established six main criteria pollutants (Shepardson et al., 2011). They include nitrogen dioxide, carbon monoxide, sulfur dioxide (SO_2), particulate matter (PM) and lead, among others. There are different areas where these gases can be produced. The EPA seeks to ensure that these gases do not affect human and animal health. Rules have been set and covered by the clean air act. Frequent analysis of the air is done to monitor the level of air quality.

5.2. SULFUR DIOXIDE (SO_2)

It is a colorless gas. The main source of SO_2 is burning fossil fuels. Industries also play a major part in the introduction of SO_2 to the atmosphere. SO_2 is known to cause severe impact to environmental health, causing significant impacts on human health. It has long term and short-term effects. For instance, if individuals suffering from asthma get exposed to air containing certain levels of SO_2 are likely to suffer from breathing difficulties. Among the short-term effects of SO_2 include causing respiratory diseases among children. These effects can be more severe in children who already have underlined conditions. These environmental impacts have been known to contribute to large numbers of respiratory cases among children and older individuals. SO_2 is known to result in the creation of secondary pollutants such as sulfate aerosols, among many others. SO_2 is known to contribute to acid rain. In most countries, sulfur can be emitted by industries but at a low amount. This has resulted in fuel standards being put in place to help control SO_2 levels in the atmosphere. Numerous measures have been put in place to ensure proper management of SO_2. It includes implementation of national fuel quality standards and promoting the use of other types of fuels.

5.3. NITROGEN DIOXIDE

It is a bad-smelling gas. There are numerous sources of nitrogen dioxide such as plants, lightning, and industries. Vehicles are known to produce large amounts of nitrogen monoxide. Other sources of nitrogen dioxide include metal refining industries, food processing industries and other manufacturing industries. There are numerous effects of nitrogen dioxide on human health. They include causing breathing complications and respiratory problems. When nitrogen dioxide reaches certain levels in the atmosphere. It is inhaled, the gas enters the lungs and rests on its walls weakening its immunity. This makes the lungs vulnerable to other infections. The case is more server among people suffering from asthma. They become more prone to suffer from more asthmatic attacks (Schuldt & Roh, 2014). High concentrations of nitrogen dioxide cause severe respiratory diseases among children and older individuals. Some measures have been put in place to ensure the release of carbon monoxide is controlled. This includes implementation of fuel quality standards, improving performance of diesel vehicles, adopting the use of alternative fuel, using other means of transport independent of fuels and development of travel programs helping individuals select better traveling options. Different countries have been put in place certain boards to control petroleum use and emission of gases.

5.4. LEAD

Lead is a major component in vehicle fuel. Lead as a natural element has several uses. It is used in making car batteries, computers, televisions, and screen protectors. When a fuel is burnt, it releases lead to the atmosphere. However, as lead is very poisonous, governments had to ensure it is removed as a petroleum product. There are numerous sources of lead emissions, including industrial emission. Lead can pollute other environmental components such as soil and water. Lead causes dire impacts on human health. Most bodies dealing with pollution have termed lead as a health hazard. When inhaled it enters the bones damaging them. Once inhaled, lead enters the blood stream and starts to accumulate in the blood stream. Depending on the frequency of the exposure, the amount of lead in will increase to a level harmful to the body. Apart from inhaling, lead can also be ingested (Singh and Purohit, 2014).

As lead is ingested or inhaled, it affects certain parts of the body such as the liver, brain, and kidneys. Toxic lead levels may cause these body parts to shut down. Lead has different impacts on different groups of individuals. For instance, in in pregnant women, lead could result in premature birth, miscarriages or impaired development in children. This is possible through the transfer of lead from the mother to the child through the placenta or breastfeeding. Lead causes significant damages among children. They may suffer from mental disability and may have a challenge learning. Some may have to repeat certain classes. Lead really affects their development, making their brains sensitive to growth. The presence of lead in children's bodies may make them have pica. This condition makes children eat non-food substances having high levels of chemical substances. For instance, the child may tend to eat paint. It also changes their feeding habits in that they can consume larger portions of food than required. When consumed at a high level, it results in chronic lead poisoning, which results in long term health complications.

5.5. OZONE

This houses photochemical oxides. This is a certain type of smog in warmer seasons. This smog can be experienced in two manners. It can either be invisible or visible as a whitish haze. These oxidants are formed when rays from the sun interact with chemical mixtures. Ozone and formaldehyde are among the main photochemical oxidants. The level of ozone in the air is used to measure the level of photochemical oxidants in the atmosphere. Most areas exhibiting high amounts of sunshine tend to exhibit high levels of photochemical oxidants. This is the same case in areas having high temperatures. Ozone is developed as a result of the interaction between other air pollutants and nitrogen oxides (NOx). This interaction results in the creation of a reactive organic substance. However, this reaction requires the presence of sunlight.

Some of the elements used involved in the reaction process are produced mostly by industries. Motor vehicles are the major source of chemical parts. Ozone gases also affects the respiratory systems. It affects the lungs and airways, among other areas. For some individuals may suffer from chest pains while others experience frequent asthma attacks.

Ground-level ozone also has some impacts on the air quality. It can significantly damage an individual's health. It causes both short term and long-term effects to an individual's health. Some of the short-term effects include swollen lungs due to the inflammation of the inner lining of the lungs, shortness of breath, and respiratory complications. Exposure to ozone makes an individual vulnerable to several respiratory diseases (Wang, 2015). Long term damages of ozone gases may cause permanent lung damage. Several steps have been taken to ensure proper management of photochemical oxidants. Some of the measures taken involve implementation of fuel quality standards, utilization of other sources of fuel, proper analysis of air quality and using proper traveling programs.

5.6. PARTICULATE MATTER (PM)

This houses a variety of substances exhibiting both chemical and physical properties discrete in nature. These substances can be in solid or liquid form dependent on the size of the particles. There are different sources of such particles. Some are derived from man-made sources, while others are obtained from stationary sources. Some may be obtained from natural sources. There are two ways in which these particles are emitted. There is direct and indirect emission of such particles (Schellnhuber, 2008). Direct emission involves these particles being released in the atmosphere individually. Indirect emission involves these particles being emitted in the environment as a result of the emission of other substances. Physical and chemical properties may vary depending on the duration, source of emission, meteorology, and area of emission. The EPA has set regulations. However, to set the regulations, the EPA has to categorize the different PMs depending on the particle sizes. Therefore, there are regulations for coarse particles and some for fine particles.

Particle matter is named according to the aerodynamic diameter. Most of these particles have a small diameter which enables these particles to easily penetrate and enter the lungs. Some of them are known as thoracic particles as they are able to penetrate the thoracic cavity. There some thoracic particles coarse in nature. They are referred to as thoracic coarse particles. Their aerodynamic diameter is larger than that of thoracic fine particles. There is a difference in the level in which coarse and fine thoracic particles penetrate the lungs. Thoracic coarse particles have large sizes that may limit the rate at which these particles enter the lungs. Some of these practices are unable to penetrate due to a number of processes (Trenberth et al., 2014).

Most thoracic coarse particles are emitted through continuous or uncontrolled burning. Some of them are produced through mechanical processes. Fine particles are produced as a result of burning processes. It can also be produced as a result of chemicals interacting with the atmosphere. With PM, the level of environmental implications is dependent on the size of the particles. For instance, fine thoracic particles are known to cause more damage compared to thoracic particles. Some of the health hazards of PM include higher numbers of premature child deaths, chronic respiratory diseases, and heart related diseases. The severity of health complications is dependent on the exposure time. There are both long- and short-term effects. Some of the complications of PM are majorly respiratory-related.

There are numerous elements composing PM. Some elements are organic in nature and can either be primary or secondary. Organic elements affect atmospherical radiation. Some of these particles have a high ability to absorb light. This kind of PM is emitted directly. Secondary organic PM is emitted indirectly. This implies that they are emitted as a result of combustion. Chemical properties of PM influence its ability to react with sunlight. It is measured by its refractive index. The refractive index determines the amount of sunlight that is scattered and the amount used in reactions.

Element of PM causing impaired vision include carbon monoxide, SO_2 and nitrogen oxide. PM is known to be hygroscopic in nature. This is due to the conversion of sulfite to sulfate. This conversion is made possible in the presence of suitable temperatures and humidity. When dealing with PM, particles are involved. These particles are deposited in different areas depending on the process. The process is dependent on the size of the particle. It is general knowledge that particles that are light and small in nature tend to stay afloat (Victor, 2011). Their size makes it possible to float in the air. The lighter the particle, the higher the level it occupies in the atmosphere. Particles that are large in size and heavy may occupy lower atmospheric levels. This is because their weight draws them towards the Earth due to gravity. The small particles stay in the atmosphere for long durations of time and are displaced through precipitation.

Different technologies are used in controlling PM. This is because there is need for regulated emission of particles in the atmosphere. This is a response to the high levels of PM in the environment. One of the measures includes installation of dust collection systems. These systems are fitted with cyclonic separators and electrostatic filters, among others. Coarse thoracic particles are removed by cyclonic separators. In cases of removing fine thoracic particles, the cyclonic separators should be modified allowing them proper removal of such particles. Cyclonic filters are beneficial in that it does not require routine maintenance practices to ensure efficient working. In some instances, industries use fabric filters.

PM has a significant impact on the environment. For instance, they affect the amount of sunlight reaching the Earth. There are both direct and indirect impacts of PM on the environment. Some of the direct impacts of PM on the environment include atmospheric radiations. This results in the generation of a radiational force. This plays a part in the amount of radiation reaching the Earth's surface. Research shows that some of these radiations spread through to space, contaminating the region. Some of the indirect effects of PM is the Cloud albedo effect. This is when the particles released in the environment interact with clouds resulting in the creation of cloud droplets. Depending on the amount of particles released in the environment the frequency in which cloud droplets are formed.

The composition of PM will influence direct and indirect effects. For instance, sulfates are known to slow down the process of global warming. It has resulted in a decline in modern climate. Soot also plays an important part in the impacts of PM in climate. Black carbon contains certain elements allowing absorption of aerosols present in the atmosphere. Most of these elements are organic in nature. PM affects the respiratory system. The size of the particle will influence the extent of PM infiltration. Large particles may move into respiratory surfaces through cilia. They may reach the throat. Smaller particles get further into the respiratory system, with some of them reaching as far as the bronchi. They can also enter the lungs penetrating the deepest parts of the lungs. For individuals having existing respiratory conditions when exposed to these particles, it may lead to server health implications (Villar and Krosnick, 2011). Smaller particles may penetrate the lungs and enter the blood stream. Some of the particles may enter other body organs. The shape of the particles will also determine the rate at which these particles enter body organs. The chemical composition of PM also affects the rate at which they penetrate into the blood and other body organs. The penetrative ability of certain PMs can be measured using a certain nomenclature. Fine thoracic particles can penetrate through to the bronchioles while some of the elements can reach the breathing system.

Some of these particles accumulate to a certain amount resulting in heart failure, among other issues. There is still a challenge with measuring the extent to which PM is hazardous. There are many health implications of PM. Some of these health implications include lung cancer, asthma, and premature death. Some of the areas where PM is causing significant damages. One of the countries is Australia. High levels of PM have been recorded in areas where coal mining is done. An increase in these levels was noted however, over a certain duration of time, it dropped as a result of the measures put in place to ensure that the levels of the PM is controlled. In China levels of PM was very high back then. However, over the years, there has been a decline in the concentration level. A body has been put in place to control air quality. This information is usually displayed in certain social media platforms (An et al., 2018).

Countries that experience high temperatures may record high levels of PM. The challenge is that high temperatures make it difficult to manage it. In countries such as Ulaanbaatar, a large part of the community makes use of natural fuels such as wood. When these resources are burnt, they produce certain pollutants such as nitrogen and SO_2. This contributes to the increased amounts of PM in the atmosphere. During cold seasons, elements of PM condense resulting in reduced visibility. In severe cases, visibility is diminished to the extent that planes are unable to land in airports. It was also discovered that ash ponds are a source of PM. PM is spread through wind erosion.

When PM is spread over water, it results in the creation of marine debris. This results in water pollution measured by evaluation of particles suspended on water surfaces. Apart from water debris, some of these particles by be released in space hence the namespace debris. The amount of particles released in the atmosphere is managed by an astronomical board. Particles released in space can be grouped into four categories. They include micro-PM, operational debris, inactive payloads, and fragmentation debris. In the case of micro-PM, as the name suggests is a matter which cannot be seen. Operational debris refers to matter produced when objects such as rockets are launched into space. Fragmented debris occurs when an object in space is broken down. This is when these projects explode or collide with other objects.

Epidemiological studies are done to evaluate the levels of particulate pollution. It is useful in categorizing PM into categories depending on their sizes, with particles having an aerodynamic diameter above 10 micrometers classified as coarse PM while those having an aerodynamic diameter below 2.5 micrometer fine particles (Al-Amin et al., 2013). Acute or short-term effects are experienced when an individual is exposed to PM over short periods of time. Long term effects are felt when individuals are exposed to PM over a short period of time. Some of the environmental implications have led the EPA to categorize PMs and establishing what concentrations are acceptable.

5.7. CARBON MONOXIDE

It is a colorless gas. It has no smell or taste. It is said to be highly toxic because when inhaled, it could lead to death. Carbon monoxide is generated through partial burning of carbon. However, the greatest producer of carbon is emissions from vehicles. When emission on carbon monoxide is not controlled, it enters the body of human beings or animals and causes reduced oxygen supply intake and supply to other body parts. CO is known to reduce the production of hemoglobin in the body. Some other physical properties of carbon include the fact that it is lighter than air. This is because it has a lighter density compared to air. This property enables carbon monoxide to readily combine with hemoglobin to form carboxyhemoglobin.

Some sources of carbon monoxide include anthropogenic emissions, emissions from cars, industrial processes, and burning of carbonaceous materials. Emissions through motor vehicles has noted an increase over the years. There are no biological and biological sources of carbon monoxide. Comparison of carbon monoxide concentrations in urban areas and rural areas show that there are higher concentrations of carbon monoxide in urban areas. This is attributed to the fact that most of the carbon monoxide found in the atmosphere is generated through vehicle emissions, and most vehicles are found in urban areas. When air samples are taken to measure carbon monoxide concentrations in traffic and other areas, high levels of carbon monoxide will be recorded in areas of high traffic. Carbon monoxide can also be generated in homes. There are home appliances using gas or kerosene. When these substances are burn, they release a certain amount of carbon monoxide in the atmosphere. Smoking of tobacco also results in the emission of carbon monoxide in the atmosphere.

Numerous methods are used in determining the amount of carbon monoxide in the air. One of the methods involve using infrared radiations. The rate of absorption of infrared radiation in a non-dispersive photometer is measured. Information on the rate of absorption will be used in estimating the amount of carbon monoxide in the air. This method is commonly used in monitoring stations. Over the years, advancement in technology enables individuals' measure in different areas by facilitating portable analysis. Another method of measurement involves electrochemical reactions. This method entails the exposure of air to certain chemicals (Zhang, 2017). Carbon monoxide present in the air will react with these chemicals to form a resultant solution indicating the presence of carbon monoxide. More tools such as sensitive passive samplers have been developed to help in easy monitoring of carbon monoxide levels in the atmosphere. Most cases of exposure are from those who live in areas where gas exhausts are the order of the day. This includes factories or those living in areas where there are vehicles. Smokers also record high levels of carbon monoxide in their blood.

There are different exposure routes for carbon monoxide in the body of a human being. The most common route of exposure is the lungs. It is more susceptible to air pollutants compared to other body parts. Other routes of infection include the skin and gestational exposure. Skin exposure involves the gas coming into contact with the skin. It is later on absorbed in the body. Gestational exposure and dermal exposures are likely to be considered as a cause of carbon monoxide poisoning. In most cases, inhalation is the main mode of exposure. Air enters the body through the nose or mouth. It enters the lungs and diffuses into the bloodstream. While in the bloodstream, it readily combines with hemoglobin to form carboxyhemoglobin. Carbon monoxide levels in the blood stream can be measured using a spectrophotometer or a suitable chromatographic method. There is certain method that measure the level of carbon monoxide toxicity in the bloodstream. When carbon monoxide enters the lungs, it enters the alveoli and capillary tubes and membranes. For pregnant women, the gas diffuses into the placental membranes. It then combines with hemoglobin. Once they combine, it causes a reduced affinity to oxygen in that less oxygen amounts are taken up compared to carbon monoxide. This is because hemoglobin has a higher carbon monoxide affinity compared to oxygen.

In as much as a large amount of carbon monoxide may be entering the body, there is a portion that leaves the body. However, if the carbon monoxide intake is higher than there will be a significant amount of carbon monoxide in the body. The process of carbon monoxide combining with hemoglobin can be reversed. The challenge is when large amounts of carbon monoxide is continuously inhaled. Removal of carbon monoxide from the blood stream is faster in an adult. Some of the health effects of carbon monoxide in a human being include brain damage, damaged cardiovascular system, and poor fetal development.

Exposure to carbon monoxide results in neurological effects where some of the brain cells get damaged due to insufficient oxygen supply. These effects may be long term or short-term depending on the amount of exposure (Bassu et al., 2014). In some instances, it may affect neurological behaviors and, in extreme cases, neurological damage. Common symptoms include headaches, vomiting, nausea, and dizziness, among others. Some individuals may get into a coma or they may collapse from time to time due to oxygen shortage.

Some of the effects of carbon monoxide in animals include carcinogenic effects. This results in animals getting tumors or cancer. This claim cannot be scientifically proven. Carcinogenic effects are also linked to mutagenic effects. Carbon monoxide also has some non-carcinogenic effects. This is when carbon monoxide builds up to high levels in the blood and directly affects the functioning of blood tissues. It also affects bodily mechanical functions in that it limits the amount of oxygen entering the blood stream. This means that body tissues may not has sufficient oxygen enabling them to function properly. This results in a condition known as hypoxia. Hemoglobin also combines with other body proteins such as myoglobin. The combination results in poor cardiac functioning. Skeletal functioning may also be impaired.

5.8. VOLATILE ORGANIC COMPOUNDS (VOC)

Volatile organic compounds (VOC) can be grouped into two main categories, namely methane and non-methane compounds. Methane compounds are mainly produced by industries and green houses. These substances are great contributors to global warming. Greenhouses produce other types of volcanic organic compounds. Methane greatly affects atmospheric temperature. They affect human health indirectly. Methane occupies the same level as carbon monoxide in the atmosphere. However, methane stays in the atmosphere for shorter durations compared to carbon monoxide. Concentrations of methane in the atmosphere have increased over the years as a result of human activities, more so agricultural activities.

The challenge of methane as an air pollutant is that it is not taken seriously as it should be. This is why researchers argue that there will be an increase in methane concentration. Some countries have taken necessary measures to control methane production. Some of these measures include the adoption of better farming systems and reduced industrial production of coal and other products that generate methane. Some of the climatic impacts of methane in the atmosphere include lack of a gaseous balance in greenhouses (Betts et al., 2011). In countries such as America, there are growing concerns over methane production and release in the atmosphere. This is because most homes make use of oil and gas products on a daily basis. This contributes to the large amounts of methane in the atmosphere, and it becomes quite difficult to control it. Another challenge is that there are frequent gas leaks releasing methane to the atmosphere. Increasing occurrence of forest worsen the situation.

Release of methane to the environment has short term effects in the environment. Future increase in global warming may be attributed to an increase in methane. Short term climate change affects an individual's health when extreme climates are experienced. This results in a higher likelihood of individuals suffering from a variety of diseases. For instance, heavy rainfall leads to flooded areas which are breeding grounds for mosquitoes. Individuals living in these areas may suffer from malaria. In most cases, when methane is released in the atmosphere, it is released alongside other toxic chemicals, namely benzene, toluene, and hydrogen sulfide. The most toxic being formaldehyde. Individuals exposed to these toxic gases are more likely to suffer from cancer, among other diseases.

Some of these gases damage internal organs and the immunity system. Most of these gases mix with rain in the atmosphere resulting in the creation of acidic rain. Companies producing methane tend to release smog deteriorating air quality. Smog results in health complications such as asthma and other related respiratory diseases. Methane also has some benefits to the environment. Methane is known to reduce the formation of ozone chemicals.

Benzene is among volatile organic chemicals. By nature, benzene is colorless, and in certain conditions, it is yellow in color. It is flammable and can easily evaporate when exposed to the sun. It is highly soluble. Benzene is a natural product produced by the environment. It can also be produced through human activities such as industrialization. It is a crucial element used in making goods such as plastics and drugs. Benzene can be released in the environment through industrial processes, emission from cars as exhaust and smoking. It can be exposed to human beings through inhalation as benzene concentrations are higher in the environment compared to the indoors (Benjamin et al., 2017). Release of hazardous wastes in the environment is one method of exposure. Some compounds found in hazardous wastes decompose releasing benzene in the environment. When benzene enters the body, it attacks the cells, making them unable to function properly. It is known to affect the bone marrow causing it to reduce the production of red blood cells in the body. This means that the body gets limited oxygen supply resulting in anemia. It can also affect the body's immunity by altering the production of white blood cells.

5.9. BIOLOGICAL AIR POLLUTANTS

Most biological pollutants are found indoors. This is because most people tend to spend most of their time indoors. Biological factors play a crucial role in environmental pollution. There are different types of biological pollutants. They occur in different shapes, sizes, and structures. They are three main categories of biological pollutants, namely molecules, biological particles, and molecules. Under microorganisms, some of the air pollutants include bacteria and fungi. Allergens and endotoxins are categorized under molecules, while viruses and pollen are categories under biological particles. Biological pollutants can severely damage an individual health. This is because exposure to such pollutants makes an individual more vulnerable to intoxications, irritations, allergies, and infections.

Most individuals suffering from asthma have gotten these diseases after coming into contact with biological pollutants when an individual is at a young age. Some may develop these complications when exposed to high amounts of these toxins as they were growing. There are numerous sources of biological pollutants. They include pollen from plants, bacteria from soil and animals, droppings from pests, excretions of saliva or skin shedding by animals, mold, viruses from animals and human beings, urine from animals and bacteria from the surrounding. Most of these contaminants are small-sized in that they can be inhaled and penetrate into body tissues (Bose, 2010). Most of these pollutants tend to settle on indoor surfaces such as furniture, walls, plants, or pets. Some of the biological particles are already present in the human body however, when their amounts increase to a higher level than required, they may damage the mat damage the body.

Molds are among the biological contaminants. There are different types of molds, namely penicillium, aspergillus, and Alternaria, among others. They are formed when microorganisms settle on a surface and develop. Once they develop, they form a stain that may take the form of a fluffy object. Certain conditions are needed to facilitate the development of molds. The environment should be moist having a humidity of about 70 to 95%. Moderate temperatures of 20 to 30°C and nutrients (Rothe, 2015). This is why molds are likely to be found in poorly ventilated rooms or and water systems. Once molds develop, they generate spores. Molds are made up of two major parts, namely mycelium and filaments. Once they are ripe, the spores are released in the environment and are inhaled. Most countries record high cases of mold-infested homes. There are different modes of exposure. They can be inhaled or come into contact with skin. When molds are released in the atmosphere, they may interact with other elements resulting in the production of chemical pollutants, namely microbial VOC.

Some biological pollutants are living organisms. Bacteria is a good example. Bacterial is said to be a single-cell living organism. They are found in the air and other elements. Bacteria is useful in ensuring normal body functions. Some are found in food and are used in fermenting of foodstuffs such as milk. It is safe to say that bacteria are safe. However, certain bacteria are not safe for human being or animals as it results in poor health (Buell, 2014). Bacteria can be found in animals, indoor surfaces, or poorly ventilated areas. Bacteria can be categorized into two groups depending on their ability to retain their color during a straining process. GRAM straining of bacteria could either be positive or negative. Positively strained bacteria have one membrane while negatively strained bacteria have two membranes. Hence there are two bacteria species. Non-pathogenic and pathogenic bacterial species.

An example of a negatively strained bacteria is Legionella pneumophila. This kind of bacteria is known to cause pneumonia. This kind of bacteria thrives in moderate conditions. Moisture is also needed to ensure its survival. This kind of bacteria is mostly found air conditioning systems. Some bacteria generate endotoxins. Endotoxins are complex molecular substances found in specific types of bacteria. Some of these molecular substances can be obtained from PM, while others are found in dead materials. Inhaled endotoxins cause numerous respiratory diseases. However, endotoxins have a positive impact in children in that when it is exposed to the immune system it boosts development and maturity. Endotoxin composition includes sugar chains and fatty acids. Cell lysis is among the major methods of releasing endotoxins in the environment. Endotoxins by nature are resistant to heat, allowing them to thrive in different environments.

Allergens are biological pollutants composed of proteins. They are obtained from rodents, pests, pets, and mold spores. When exposed to human beings, allergens can irritate the respiratory system. Allergens cause diseases such as asthma. Most allergens are found in dust. Indoor allergens include D. farinae and Dermatophagoides. Allergens thrive in moderate temperatures and require enough nutrients to ensure its survival. Allergens from pests are obtained when they sweat or secrete saliva. Viruses are dangerous biological pollutants. They are infectious agent that have the ability to multiply inside the body. Viruses can damage an individual's immunity (Zaval et al., 2014). Viruses can be found in every ecosystem. Once a virus enters the body, it replicates and infects other cells. Viruses exist in different shapes and sizes. Some viruses exist in helical or icosahedral shapes. Shapes vary according to the complexity of the virus. Viruses cannot be seen by the naked eye. Viruses can be transmitted through disease-bearing organisms. These organisms are referred to as vectors. Some viruses are transmitted by plants, while others are transmitted by insects.

Viruses reproduce by replicating. Viruses are small in size, enabling them to penetrate internal body organs. Viruses have different genomes and can be categorized depending on the properties. One of the properties is nucleic acid. In terms of shape, they can be categorized as either segmented, linear, or circular. Sense is also a property that categorizes viruses either as a positive sense, ambisense or negative sense. Some viruses undergo mutations making it difficult to control them. They become resistant to drugs used incurring viral diseases. Viruses cause numerous diseases, among them chickenpox, sores, and influenza, among others. Viruses are usually transmitted by air. Some viruses require a moist environment to survive. Viruses have the ability to survive in different environments. Usually, viruses, when released in the atmosphere settle on indoor surfaces. While on the surfaces they can survive for a certain duration of time. Examples of viruses include chickenpox, influenza, and measles, among others.

Some viruses are produced by flowers. A good example is pollens. Pollens are the result of plant reproduction. People come into contact with pollen either indoors or outdoors. They are transmitted through agents known as pollination agents. These agents include wind and water. Other animals can also be used in transmitting these diseases. Some pollen act as allergens as they carry some allergens on their surface. The presence of allergens is dependent on the type of plant species. Most pollen concentration is found outdoors. General effects of viruses on human health include triggering allergic reactions such as asthma and pneumonitis. It also causes diseases such as influenza and measles. Some biological pollutants like molds produce disease causing toxins.

Depending on the level of exposure, an individual may suffer from frequent asthma attacks. Some symptoms exhibited by individuals exposed to biological pollutants include fever, coughing, sneezing, and dizziness. The body may react to biological pollinators immediately or after continuous exposures. Some individuals may get mild allergic reaction, while others experience frequent allergic reactions. Some of the biological pollutants may be advantageous in child growth. However, some children may suffer from respiratory diseases (Brulle et al., 2012). Numerous steps can be taken to reduce exposure to biological pollutants. One needs to ensure their environment is well maintained. There should be proper ventilation. For pollutants such as molds, the environment should be dried. Exhaust fans should be installed in closed areas to reduce incidence of moisture. Closed areas should be properly ventilated to reduce incidence of moisture build-up. Rugs and carpets should be properly cleaned and dried to reduce dust and moisture. Routine pest control should be done to reduce pests. By controlling pests, it reduces the spread of biological pollutants.

5.10. AIR TOXINS

These pollutants are toxic in nature in that they can cause severe damages to the life of human beings. Air toxins can be in gaseous in nature or solids in nature. They are also referred to as hazardous toxins. They are the most dangerous kinds of toxins. Some of the effects of these pollutants include them causing cancer, retarded growth and development in children, and respiratory diseases. These toxins exist in different forms. Some of them are particulate in nature, while others are found in other substances. Air toxins are categorized according to their characteristics. For instance, mercury is categorized under bio-accumulative toxins. The term bio-accumulative is used to refer to toxins that accumulate an individual's body gradually as they are exposed to it. The body becomes a storage facility. There are different modes of exposure to these toxins, such as ingestion and inhalation. Ingestion occurs when this toxin rest on food surfaces and are consumed. They can also be exposed to a human being indirectly. This is when an animal inhales these toxins over time and are consumed. Some air toxins include naphthalene, mercury, chromium, and cadmium.

Naphthalene exists as a white solid or a gas. In gaseous form, its smell is not pleasant. Naphthalene is commonly found near industries and areas having several vehicles. It is used in manufacturing several commodities such as insecticides, herbicides, and surfactants. Naphthalene has a complex chemical structure hence referred to as a special compound. It is formed as a result of the combination of two or more benzene compounds. This contributes to naphthalene being flammable. Naphthalene is said to be a non-volatile organic compound. This is because some of its qualities do not meet the standards needed for it to be classified as a volatile organic compound. However, among polycyclic aromatic hydrocarbons, naphthalene is highly volatile (D'Amato and Akdis, 2020).

As an air toxin, naphthalene has received little attention. This is alarming as many individuals are exposed to naphthalene while at work or in their homes. People working in industries are more likely to have high levels of naphthalene in their system. Numerous health complications are linked to naphthalene exposures. Nasal tumors and respiratory complications are among these complications. Studies have been done to evaluate some other health complications linked to naphthalene. This led to the establishment of a program studying chronic inhalation of naphthalene. Some countries have set up research committees to conduct further analysis. Findings for the studies have led to naphthalene being classified as a hazardous air pollutant. Activities dealing with naphthalene need to be in accordance with the EPA's regulations on hazardous air pollutants. Some requirements include frequent analysis of air in industries heavily producing naphthalene, allowing easy monitoring of naphthalene concentration in the atmosphere.

Mercury is also another hazardous toxin. Usually, it is found underground and released to the air through certain human activities. Once released in the air, mercury contaminates other environmental elements such as water and air. They combine with other elements forming highly toxic substances. It is bio-accumulative. This is proven by the traces of mercury found in people's bloodstreams. There are both natural and artificial sources of mercury. Natural sources include oceanic emission and eruptions from volcanos. Artificial sources include combustion of substances releasing mercury in the atmosphere. Artificial sources account for the large percentage of mercury in the atmosphere. Examples of activities emitting mercury in the environment include burning of coal, industrial production of cement, cremation, and combustion of metals. It has been noted that there is an increase in mercury levels globally. Studies shows that mercury is re-emitted in the atmosphere through human activities.

There are various applications of mercury. It is used in the medical industry in the production of medicines. It has also been used in making tools such as thermometers used in different fields. Mercury is majorly exported by Europe hence high rates of mercury pollution noted in the country. Countries such as Denmark and Sweden have prohibited the use of mercury in making such instruments or production of drugs. Some of the countries have implemented policies requiring hospitals to purchase mercury-free products. This has encouraged other countries to minimize utilization of mercury. This has resulted in reduced rates of mercury pollution. Eliminating usage of mercury required the presence of alternatives to mercury. For instance, mercury-based thermometers could be replaced with electronic thermometers. The challenge of using electronic thermometers is the fact that they are expensive. In the case of sphygmomanometers, aneroid kind should be used. In the case of industries, instead of mining and processing coal, options such as renewable sources of energy could be used (Dai,

2011). Industries could also use coal having less amount of mercury. Plants should use technology less dependent on mercury usage. Industries should minimize production of mercury-based products.

The medical sector should look into production of vaccines containing little or no mercury. This may require availability of refrigerators in hospitals. This is because mercury is mainly used as a preservative. Some of the health effects of mercury include distorted vision, impaired hearing, poor development, and impaired speech. Among children, exposure to mercury causes poor brain development, poor memory, impaired learning, and low levels of concentration. In some cases, it results in a disease known as Minamata disease. This is when the child is exposed to high levels of mercury. Among adults, it causes deformation in limbs, reduced intellectual ability and salivating more than usual. Among pregnant women, exposure to mercury mainly affects the developing fetus. It causes retarded growth and poor mental growth.

Mercury pollution also affects animal health. Research shows that extreme exposure to mercury results in behavioral changes in animals. It also affects their mortality rates and development. Some of the symptoms of animals exposed to mercury include animals collapsing. This was witnesses when numerous birds fell from the sky after they collapsed dead. In fish, it was noted that those who lived in infected water tend to swim in circles.

Cadmium is a soft white metal. It is a stable chemical element. It usually does not exist singly. It is commonly found in zinc ores. Cadmium is highly toxic, resulting in a decline in its use. It has been identified by the EPA as a hazardous pollutant. Cadmium is known to be ductile and has a high resistance to corrosion hence used in protecting highly corrosive elements. It has a low solubility rate in water. It becomes flammable when combined with other elements. When ignited, it burns with a dark brown color. It is a naturally occurring material though rare. It is found as a zinc impurity and can be found in industries producing zinc. Cadmium is used in making batteries more so those that can be recharged. It was used in making the positive electrode. Examples of batteries produced using cadmium include silver cadmium battery. It has also been used in electroplating metals such as steel. However, using cadmium in electroplating steel presented a challenge; hydrogen embrittlement. This required numerous processes to be done before cadmium is used (Diffenbaugh and Burke, 2019).

Cadmium is also used in nuclear plants. They are used in making control rods useful in controlling nuclear reactors. They have the ability to absorb neutrons reducing chances of other fission reactions from occurring, facilitating a controlled reaction. It has also been used in making televisions and producing anticancer drugs. Scientists have considered using cadmium as a cancer treatment option. The challenge is removing toxicity in cadmium. It is used in laboratories in making lasers. It involves combining cadmium and lasers. Cadmium is known to produce some kind of illumination when excited by ultraviolet radiation. They can be used in the medical field in the process of creating tissue images. Some of the health risks presented by cadmium are mostly toxin-related. They can contaminate the body and the bloodstream. Like cadmium, chromium is also a chemical element. It is a transitional metal, usually gray in color. There are several chromium alloys including ferrochromium. Chromium is highly reactive and has a high resistance to corrosion. Like cadmium, it has been used in coating highly reactive elements. Electroplating metal using chromium is referred to as chrome plating. Its ion is of high importance in the body as it enables the production of sugars and insulin in the body. Chromium is soluble in water and can be reduced, resulting in the formation of other products. It can also react with other chemicals resulting in the formation of chromium sulfates and chlorides. When oxidized, chromium can occupy different oxidation states. Chromium is extracted from chromite ores. There are series of reactions used in producing pure chromium. Some of the areas where chromium is used include strengthening of metals used in making normal utensils. Chromium is used in coating steel resulting in the creation of stainless steel. Physical's properties such as the hardness of the material is used in covering surfaces. Its hardness makes chromium to be considered as a better material to be used in electroplating of metals. The method of electroplating will be dependent on whether the coating should be thick or thin. It is also used in pigmentation and making of corundum crystals. Chromium is still a major health hazard and is listed under hazardous air pollutants.

5.11. ENVIRONMENTAL EFFECTS OF AIR POLLUTANTS

Apart from human health, air pollutants have severely affected the environment. One of the environmental impacts include the creation of acid rain. Acid rain is formed when certain air pollutants such as carbon monoxide combine with rain. Acid rain can reach the earth surface as fog, snow, rain, and in some cases as dry precipitation (Donnelly et al., 2017). Acid rain causes significant damages in plants and animals. For instance, most of these acid rain accumulates in rivers and lake making them toxic. These toxins are later on ingested by animals and plants, damaging their health. It has also resulted in increased acidity in soil, making it unsuitable for insects to live there.

Air pollution has also resulted in eutrophication. Eutrophication refers to increased concentrations of certain nutrients in ecosystems making the environment unconducive for plants and animals. When these gases are released in the atmosphere, they settle in water bodies, increasing their concentrations. Eutrophication is liked to increased concentrations of nitrogen in the atmosphere. Eutrophication caused death in aquatic life and negatively affects plant biodiversity. Increased air pollution increases eutrophication more so in aquatic ecosystems. Air pollutants have negatively affected wild life. For instance, in fish, some corrosive air pollutants rest on their skin and corrode their skin. In other aquatic animals such as frogs, particles from these pollutants rest on their skin, blocking their breathing. The same case applies for plants. Particles from air pollutants clog the stomata hindering entry and removal of oxygen and carbon dioxide from the plant. This hinders proper growth and development in plants. Some biological increase the incidence of an animal suffering from respirational diseases. Toxic air pollutant accumulates in the body of an animal making them susceptible to other diseases.

Release of air pollutants in the atmosphere has led to the depletion of the ozone layer. The ozone layer is useful in prevent harmful rays from reaching the earth surface. Air pollutants contain chemicals which react with the ozone layer causing its depletion. Depletion of the ozone layer allows entry of harmful sun rays on the Earth's surface increases cases of skin cancer, among other diseases. These harmful sun rays also damage crops highly sensitive to ultraviolet radiation. Pollutants such as ground-level ozone have contributed to increased numbers of damaged forests. This pollutant affects plant growth, thereby affecting its yields. Some of the pollutants have resulted in reduced visibility due to the production of smog. As mentioned earlier, some of these pollutants react with sunlight resulting in the formation of colored substances that diminish visibility. This has resulted in increased road accidents.The greatest impact of air pollutants on the environment is climate change. Emissions of gases such as nitrogen oxide and carbon monoxide in the environment disturb the natural balance of gases in the atmosphere. This disturbs natural cycles useful in controlling the Earth's temperature. Increased amounts of gases such as carbon monoxide in the environment have resulted in global warming.

Chapter

6

Solutions to Air Pollution

CONTENTS

6.1. INTRODUCTION

When developing a new air quality policy, several factors must be considered, including current initiatives and partners, as well as how to branch out to key stakeholders for wider consultation. Some of the steps include:

6.1.1. Reviewing Existing Planning Documents for Solutions That Have Been Identified

The majority of national governments have National Action Plans and other types of national plan documents, which can provide insights into the objectives for action of this government. Related documents may be available to local governments, in particular local transportation or health departments. If the documents are not yet available, some initial planning in accordance with the Megacity Framework might be helpful through using available knowledge and enforcing a few measures that are feasible. Imperfect data still are adequate for most cities to start planning with the effort to achieving data over time (Figure 6.1) (Hiltner, 2009).

Figure 6.1: Air pollution is mostly replaced with alternative energy sources such as solar, wind, and geothermal fuels. Fossil fuels are the primary solution. It is essential to produce clean energy. But our energy consumption must also be reduced by using responsible practices and more efficient equipment.

Source: https://www.unicef.org/vietnam/stories/making-every-breath-count.

6.1.2. Identifying Local Partners

In addition to local and national governments, a large number of international air pollution groups have made significant investments in the area of health: the UN Environment Program, the Clean Cooking Alliance, the World Health Organization, and the World Bank. The use of sustainable optimization and local stakeholders often results in clean air. Models can be developed for decision-making at a city-level level. Essential local partners may also include educational institutions that have the technical skills necessary for government support or can be in a position to purchase them (e.g., modeling and monitoring).

6.1.3. Investment Deeply Rather Than Broadly

Initial investments may include support for mitigation of priorities identified in current planning documentation. The early work could be focused on one or two sectors, such as home-cooking (e.g.) or domestic industrial (e.g., enhanced brick ovens) or urban and civic issues (e.g., reduced open trash burning). Further investments could focus on better monitoring of air quality; data processing and training; extension and communication planning; secondary town surveillance. Future work could include emission standards for industrial sources; laboratory capacity building analysis and training, and enhanced vehicle standards.

6.1.4. Development of a Communication Strategy

Communication and community engagement strategies are essential for all air quality programs. From the early planning stages, stakeholder groups should be regarded. While many low-income residents fail to fully understand the relationships between air quality and clinical outcomes or the sources that most contribute to the local air pollution, the affected public may be one of the strongest motivators of political will. Public information campaigns, community health interventions, medical professionals, and religious and traditional representatives in many places can be the main focus of investment efforts. Communication methods have included: who listens to the public? Where would they get trustworthy information? Awareness of high air pollution levels is critical for building growing outrage and interest (Schuldt and Roh, 2014). In advocacy, it would be important to make air pollution more visible. The danger of air pollution to the health can also be efficient in interacting and can reach a wider audience, including the health sector. Several countries developed Air Quality Indexes that relate color densities to suggestions on which demographic subgroups should take measures to limit exposure. The WHO BreatheLife Campaign is developing additional communication tools.

In addition to local interaction, world-class high-profile events (e.g., a special meeting on air quality at the UN General Assembly) receive a unique opportunity to link international air pollution participation with the country's involvement in key matters. Highly burdensome countries in Africa and South Asia could play a leading role with a certain amount of advance planning. If multi-sectoral involvement is expanded throughout the continuum of air pollution, then cross-border long-range air issues are also important for world debate. For more communicating effectively with the public and policymakers in particular locations, available products can be tailored to the local framework. Officially, the towns of India are covered in the Air Pollution Knowledge Assessment Program (ApnA), which extends to other regions and shows how regionalized data can be viewed for the unique level of personal cities. As both air pollution and greenhouse gases (GHGs) are often produced by emission sources, raising public awareness can help mitigate these issues.

The multi-sectoral process requires a range of stakeholders to promote and impact initiatives on air pollution through financing, advocacy, power generation, workable ideas, and much more. Some of the parties concerned include:

- Children and school;
- Communicators including media;
- Diaspora;
- Governments that include Ministry of Finance that can levy taxes as incentives;
- Private sector that includes corporate social responsibility, private companies driven by profit, and private-public partnership;
- Civil society organizations and communities that are affected;
- Professionals in city management, health, and environment (transportation and energy);
- Professionals in public health (inclusive of health sector managers, Ministry of health);
- Academic institutions;
- Extension workers, health providers, and their medical institutions.

6.2. ACTIONS FOR REDUCING LEADING SOURCES OF AIR POLLUTION

This chapter highlights pragmatic approaches in LMIC environments to overcome pollution challenges in the form of residence cooking with cleaner fuels, relatively clean manufacturing technologies, cleaner vehicles and fuels, reduced farm burning. It is important for better air quality to deal with regional sources beyond urban areas (Figure 6.2).

Figure 6.2: The effects of air pollution are still evident, although laws and regulations contribute. The decrease in toxic air and improvements in health are welcome changes. But there remains a challenge which needs to be addressed by the growing danger of climate change because of emissions of fossil fuel.

Source: https://www.pinterest.com/pin/629096641683248778/.

6.2.1. Transitions to Clean Household Fuels

The ineffective combustion of solid cooking and heating fuels in household air pollution (HAP) continues to be a major global health concern and is a major contributor to environmental pollution. HAP exposure is anticipated to reach to several million premature deaths in LMICs every year and includes no additional health effects from household instrumentation air pollution, which has been demonstrated to reach 25% of Indian AAPs. Although efforts to reduce targeted HAP cleaner burners were made at an early stage, recent attention stressed the need to ensure cleaner fuels are available to their maximum benefit in health. The change into clean household fuel, including biogas, LPG, ethanol, cooking, and pellets electricity, can reduce HAP attention and improve maternal and child health (Rothe, 2015).

Early proof has demonstrated that decreasing publicity to HAP can reduce adverse health outcomes. Burning chimneys was demonstrated in Guatemala and Mexico when the use of chimney chambers and closed-burning chambers decreased the HAP in serious cases of pneumonia, relative to kids in households where wood was opened.

In comparison to firewood or kerosene furnaces, randomized controlled trials (RCTs) in Nigeria have found use by ethanol cookstoves improve birth results, such as higher birth weight and less neonatal death, lower maternal blood pressure, and reduced inflammatory blood biomarkers (BBIs) in females who are pregnant.

These research results show the potential health benefits which can help lower HAP exposure. Reducing HAP's health impacts, however, means reducing dramatically population exposure to HAP, as set out in WHO Indoor Air Quality Guidelines, and cannot be easily accomplished without moving the biomass away in the present format (dung, crop residues, charcoal, and firewood).

For decades better biomass cookware have been promoted for optimum efficiency but not necessarily clean emission, and while often effective in laboratory tests, the WHO air quality guideline for exposure to particulate matter (PM) does not in practice apply. Clean household air needs to be replaced by clean household sources of energy, like biomass, ethanol, gas, and energy, gas from biomass burning. It promises to make the transition from households to clean indoor air hurry if households have access to cost-effective and accessible clean fuels. We summarize each fuel option briefly in this section. Piped natural gas (PNG) and biogas liquefied petroleum gas (LPG). LPG is a clean and portable fuel which is a by-product of the production and refining of oil and natural gas. There is an extensive LPG source; there is a global LPG surplus and it is expected to remain in 2050.

PNG permits the availability of consistent gas in the household. In addition, the fermentation of organic waste and residue can be used to produce renewable biogas. As biogas, PNG, and LPG, combustion emit a significant reduction in PM, LPG, PNG, and biogas fuel emits less air pollution than burning wood, carbon or kerosene and the health burden associated there with them is expected to be reduced. In addition, biogas, PNG, and LPG combustion lead to the emission of much less black carbon in comparison to solid fuels, a major agent for global warming. If countries incorporate and implement sustainable, well-established models for expansion and safety in the gas field and invest in necessary infrastructure with suitable expert assistance, the national availability, adoption, and use of LPG may be rapidly expanded (Robinson et al., 2009). A holistic approach is essential: programs and policies should make savings, safe, consumer-friendly LPG stoves, the proper layout of conditions for a sustainable, bankable LPG industry, along with the supply chain for functional fuel, cost, and competition rules and an environmental safety ecosystem.

Countries such as Brazil, Malaysia, Indonesia, Côte d'Ivoire, Senegal, Senegal, Malaysia, India, and others have made use of LPG cooking in 75% or more and sometimes in more than 50% of the countryside. In many Latin American countries, LPG already is the major cooking fuel, and the LPG use is rapidly expanding in India and Indonesia. The transition towards LPG may be tackled in certain ways by micro-finance and household subsidies. Biogas can be regarded as clean fuel 'shop.' A minimum of animal waste and considerable domestic effort as well as appropriate temperature specifications for the continued production of gas are required for the production of biogas. In certain rural and peri-urban areas, for instance, in which adequate animal waste is easily obtainable, biogas programs have succeeded. In Nepal, biogas is a responsive clean fuel option for more than a decade, but in the winter months, households typically need backup energy. To achieve true clean household energy, it is essential to ensure the availability and use of clean additional sources of energy, like electricity or GPS.

1. **Ethanol Fuel:** Ethanol is a renewable biofuel, which can be used to cook and is cleaner than traditional fuel for biomass at a cost that, depending on geographical location, is similar to that of kerosene and charcoal. Ethanol cooking has the ability to reduce HAP exposures efficiently and to increase health outcomes. In

2017, a clean cooking methods meta-analysis demonstrated an 83% reduction in emissions for ethanol stoves. When plants are utilized for the stock of biofuel, competition may take place for land used for food production. The major challenge for the wider use of ethanol as a clean household fuel has been to ensure constant supply, especially in the transportation sector. The supply chain of ethanol is largely worldwide. The distribution of the ethanol in the refillable canister has been addressed by Koko Networks operating in East Africa and India (Revesz et al., 2014). The molass waste stream offers a sustainable supply of feedstock for the production of ethanol, which is a far more sustainable source than that of maize and other foods diverted to the production of ethanol (Figure 6.3).

Figure 6.3: There is a need to request countries to enforce strict regulations requiring air scrubbers in all industrial installations that release pollutants from the atmosphere.

Source: http://web.mit.edu/12.000/www/m2015/2015/airpollutionsolutions.html.

2. **Pellets:** When produced and adopted with the suitable source materials and by households, biomass pellets and briquets can become clean fuel for the future. Research from the biomass

evaluation in Kenya and elsewhere shows adequate subject matter in volumes that can meet the demands of household cooking to generate pellets-based fuel. Pellets are manufactured to produce a coherent and clean-burner material through compression of biomass. Clean domestic air is only achieved when pellets are produced to a high standard, burned to the emission targets of the Tier 4 stoves, and burned in appropriate stoves, including stove repairs and pellet storage, properly operated. Although pellets can be manufactured from biomass in most areas, the raw materials must be dry, accessible, and compressed year-round using specialized equipment. It is rewarding to produce consistent pellets of high quality with a supply chain capable of supplying pellets all seasons long. However, scaling pellets on a sustainable market basis, itself is a challenge and requires a high-quality assurance, as well as a business model for household pellets, now serving over 15,000 households, Inyenyeri, a hybrid company based in Rwanda (Robinson et al., 2009).

6.2.2. Using Cleaner Brick Kilns to Reduce Community Air Pollution

Brick production in many LMICs in South Asia is a key source of ambient atmospheric pollution, with some 300,000 brick kilns contributing to air pollution in the region. Reducing brick kiln emissions has the potential to improve air quality significantly. In the early winter to last spring in Nepal, around 1000 brick ovens operate and contribute 40% to the area's basic carbon. In the valley, the earthquake in Kathmandu in 2015 damaged 110 of the 112 brick kilns, and the damage done to brick stoves was a unique chance for energy-efficient and less polluting brick kilns to be rebuilt. A partnership has united to develop new technologies for the production of bricks, reducing emissions, using energy, and improving the quality of brick. One result was the reconstruction of 'zig-zags' fixed brick chimney furnaces that are resistant to earthquakes, energy-efficient, and worker-friendly and has reduced emissions (Hiller, 2011).

Under ICIMOD leadership, training, and site visits have been coordinated since 2015 to promote transformation from the fixed chimney bulls (FCBT K) trench kilns to zig-zag kilns for brick entrepreneurs in Bangladesh, India, Pakistan, and Nepal. A study in Nepal found that incorporating FCBTK to zig-zag kilns is likely to reduce $PM_{2.5}$ emissions by ~20% and BC emissions by ~30%, based on the estimated emission factor per kilogram of fuel. Political changes backed the effort to transform zig-zag kilns from FCBTK. The Punjab Province (India) EPA issued a directive on zig-zag technology converting all kilns in 2018. In 2018, a new directive was adopted. In India, FCBTK has been converted into zig-zag in regions where air pollution is high. In India, the Central Pollution Control Board. In Nepal, modifications to emission standards for brick kilns were carried out on the basis of measurement campaigns supported by CCAC (Revesz et al., 2014). The formation of the South Asia Brick Kiln Association (FABKA), made up of members of India, Pakistan, Bangladesh, and Nepal, has been one of the major steps for this region. FABKA aims to facilitate cross-border knowledge sharing and work to reduce pollution and make kilns more sustainable. The change from FCBTK to zig-zag ovens resulted in lower fuel consumption, an increase in grade A and lower capital-intensive ovens that made brick entrepreneurs economically viable. In the past, technical expertise was most often transferred from father to son, and many firms lacked the information necessary for best practices or standards. Training and instruction manuals for the construction and operation of zig-zag ovens have made it easier to adopt. This achievement in the region shows how important cooperation between stakeholders is for an effective solution to be developed.

6.2.3. Reduction of Community Air Pollution Through the Reduction of Trash Burning

Significant emissions of particles, volatile organic compounds (VOCs), and a broad range of toxic pollutants, including cancer-causing dioxins, result from burning waste, particularly in urban regions. It is estimated that the reduction in waste burning in urban LMICs will reduce air pollution by up to 40%. The absence of systematic waste collection causes open waste burning in the towns of LMIC. Due to the multifaceted problem of trash disposal, including large sites, small informal dumps and incineration in the area, regulation is difficult. Burning trash reduces the waste volume and thus provides more space in waste disposal sites, but combustion also causes high air pollution, including toxic and cancerous gasses emitted from combustible plastics.

Climate and Clean Air have created a waste initiative to address the challenge of solid waste combustion. It supports policies and actions to reduce black carbon emissions from open waste consumption by national and local governments, minimize the total pollution that endangers local communities and promote universal collection by cooperating and playing an important role. The waste initiative provides partners with tools and technical resources to help decrease open waste combustion by:

- Promoting the development of integrated systems of solid waste management which enhance waste collection and reduce the occurrence of trash in streets or informal dumps likely to be opened.
- Support better management of waste disposal systems, which can reduce spontaneous fires.
- Exercise community-wide efforts to encourage waste reduction, including reuse, recycling, and composting, and educate public opinion on the risks posed by open waste incineration.

In 2018, the Ministry of Sanitation of Ghana prohibited open combustion in Accra City. In many countries, plastic sackings leading to plastic waste have been banned.

6.2.4. Movement From Open Burning (Peat Burning, Forest, and Agricultural)

Open forest burning and/or crop residue for many LMICs is an integral component of traditional farming operations. Agricultural fire emissions contribute significantly to regional air pollution.

For example, Indonesia's 6B dollar palm oil industry is responsible for the huge fires that spread too many of Indonesia, extending into Singapore in peat bogs, which create a toxic haze.

An estimated 100,000 early fatalities, 90% in Indonesia and many thousands more in Malaysia, Singapore, and other countries in the region were caused by forest fires in 2015 (Pleijel, 2009). Open burning of agricultural remains in Northern India is a seasonal source of air pollution that affects both air quality and public health in parts of Pakistan and Bangladesh. Air quality may be greatly improved by regional cooperation and action (Figure 6.4).

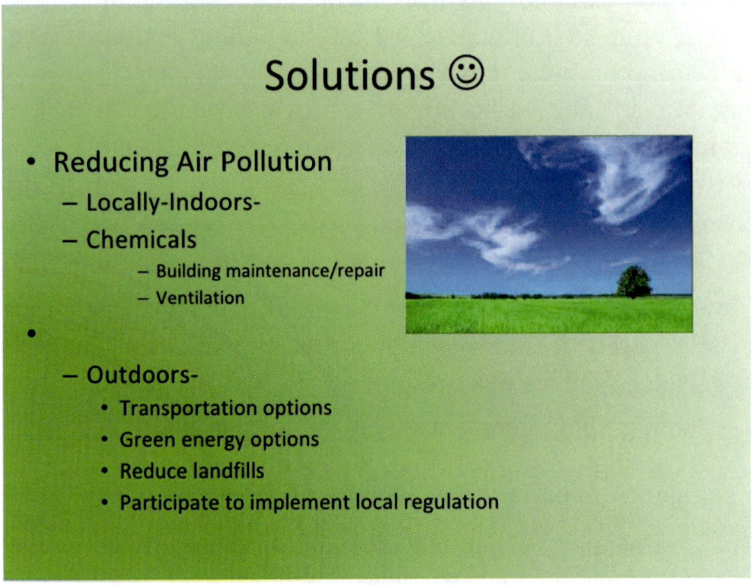

Figure 6.4: Pollution is a big term now used by our ears. Every day, we hear about and read through the mass media about the various forms of pollution. Air pollution is a form of air pollution, regardless of indoor or outdoor air contamination.

Source: https://www.slideshare.net/tastegaia/pollutions-and-solutions.

Open burning reductions are essential in downwind cities to achieve clean air. USAID works with the government and several international partners in Indonesia to encourage the recovery of peat and fight against forest burning.

The IFSC works with local communities to understand the political, social, and economic dynamics causing forest and land fires. CIFOR also investigates how land clearance and development that are not dependent on the use of fire can be facilitated and maintained. In Africa, there has to be improved management of forests and maintenance practices which address the challenges of rural livelihood and land use. In many parts of Africa, charcoal is a major cooking fuel, especially in cities, which make its compact shape easier than wood to store and deliver. As the production of charcoal, known for its rapid deterioration of forest resources, is connected with food and livelihoods, it is necessary to make the transition from charcoal a transformation of livelihoods (Jang and Hart, 2015). Charcoal is a major source of PM, including both black and organic carbon. This is low oxygen combustion in wood. Reduced dependence on carbon as a cooker will depend both on alternative domestic fuels and on new jobs, including the delivery and maintenance of cleaner fuel and household technology.

6.2.5. Focus on Innovations Reducing Transportation Sector Sources

In LMIC towns, including in Africa, the transport sector is an increasingly important source of air pollution where older, poorly tuned vehicles using a variety of fuels could lead to high emissions and localized impacts. Increasing traffic congealment, weak regulation and lack of enforcement exacerbate the problem. Important steps were made in the elimination of gasoline lead in all African countries and the reduction of the sulfur content of diesel fuels, allowing high efficiency exhaust particulate traps to be used in some countries.

Additional entry points for emissions mitigation from cars include improvements to fuel standards (specify carbon and sulfur and commercial fuel quality), regulations requiring catalytic vehicle emission checks (which will work as planned only if fuel quality is checked), and inspection and maintenance programs requiring annual inspections and tests.

The natural starting point for transport systems to mitigate the emissions of air pollution in LMIC cities is likely a mixture of one or more of these solutions. Innovation in e-mobility can lead the way towards transport and lead by emerging developments in countries with high incomes. A new "self-supplied, shared, and electric" paradigm is emerging within the United States. A related but minor variation, namely "connected, shared, and electrical," has been proposed in India.

Increasing by 30% each year (in the world) and dozens of available models for electric vehicles, market indicators for growth and innovation have been demonstrated each year. Governments and automobile companies' commitments drive market and policy development.

The Indian government committed $1.4 trillion to electrify transport. Ford Motor Company plans to provide $11 trillion for the development of electric vehicles with 40 electric vehicles promised by 2022. It is a major question when and how these innovations are to be accessible in LMIC cities (Kenis & Mathijs, 2012).

Innovations in e-mobility have the potential to influence local air pollution and the mitigation of climate change. In order to achieve profound economic decarbonization, renewable electrification for the transport sector is an important component of the various paths. Transport, for example, into LMICs represents a relatively small portion of the fine particulate emission ($PM_{2.5}$) on a domestic level (2%); on a local level, however, transportation accounts for around 20–30% of the emissions of $PM_{2.5}$.

In India, the relevant trends include the increase in the energy load (and revenue sources that can help combat the financial problems of utilities) and the integration of renewable energy sources, as electric vehicles offer a battery source. Another important factor is the storage capacity of electric car batteries, and other storage will dwarf (Kenis and Mathijs, 2012). A national electric mobility mission has begun in India.

It covers all segments of the value chain of e-mobility: manufacturers, charging infrastructure firms, fleet operators and service providers. These policies include public procurement of the fleet, readiness to use electricity equipment, financial incentives (and incentives to polluting vehicles), bulk procurement to reduce costs and an overall target of a 30% EV penetration by 2030.

The main challenges are driving range anxiety, cost, and lack of consumer awareness; inadequate charging infrastructure; disagreement over the role of utilities in the financing, ownership, and operation of the EV charge infrastructure and the lack of interoperability standards that have been developed; and distribution network concerns regarding the impact of charge. On all these fronts, progress is being made. The range of travel has considerably increased and reaches 400–500 km. Capital costs decrease, and within a few years, some projections estimate parity.

6.2.6. Monitoring Levels of Air Pollution

A first important step in understanding the relevance of air pollution in all locations in the monitoring of air pollution levels. Air quality control enables a number of important objectives to be met: air quality assessment and health risk assessment, attribution of the source, compliance, and accountability understanding of the contribution of component parts of air pollution, trend monitoring, and public information (Kump, 2011). Surveillance can also be used for atmospheric research and the evaluation of models of exposure (Figure 6.5).

Figure 6.5: Governments, councils, and planning bodies now have to begin to include measurements of air quality in their strategic plans and planning decisions. It is probably negligent or irresponsible to deliver or to encourage development knowingly in areas with poor air quality (or the risk).

Source: https://www.manxtechgroup.com/monitoring-air-pollution-with-iot/.

In numerous countries, ground-based air pollution surveillance networks operate. These networks comprise one or more stations in cities and other locations strategically situated (e.g., Addis Ababa, Delhi, etc.). Climate-controlled structures may consist of stations which have several high-quality measuring instruments such as CO, NO_2, SO_2, PM_{10}, and $PM_{2.5}$, and possible others. Many of these measurements are collected and shown on www.openaq.org in real-time for the viewing of observed levels of air pollution. Because this type of high-quality air monitoring system is cost and technically required, many LMICs lack ground-based monitoring systems.

Lack of observations of air pollution is an important challenge in tackling the issue of air pollution, as current levels of air pollution at these locations remain lower than at locations with surface monitoring systems. More and more sophisticated atmospheric pollution models based on remote satellite sensing, earth observations, and chemicals models enable global estimates of air pollution levels. These estimates may be used to reliably identify urban areas where pollution levels are well above health-based standards, albeit less accurate than land monitoring (Lineman et al., 2015).

Several pollutants can be monitored, but measurements of $PM_{2.5}$ are typically a good way to start an air monitoring program. A network of $PM_{2.5}$ monitor systems based on the ground can serve as a good starting point. A consistent schedule works. A network can produce granular specificities of small areas with high time resolution, depending on the equipment and site chosen. Additional information on air pollution level spatial variability may be developed through the inclusion of information for other monitor types, like remote satellite sensing, a temporary low-cost sensor network, and/or mobile surveillance. These additional sources of information can be cheaper than high quality.

Steps to select the type of network monitoring:

1. **The System Scope:** The program for monitoring the air quality ranges from simple neighborhood data to advanced surface surveillance stations that measure both levels and particulate composition for identification by the source (i.e., high air pollution day) (e.g., reporting temporal and spatial variability with high precision). Strategy for air quality monitoring and selection of sensors should be chosen to fit.
2. **Filter-based vs. Real-Time PM Sampling:** Relatively cheap real-time sensors focus on providing continuous data but do not allow accurate analysis and sampling; the integrated filter sample gives a more precise $PM_{2.5}$ measure and supports the verification of the particle composition, while offering only one sample value, which is usually done once a day.
3. **Species of Chemicals:** Many other air quality monitors report $PM_{2.5}$; further monitoring is usually necessary to measure NOx, SOx, CO, and other chemical species. The $PM_{2.5}$ mass monitoring is a critical initial step, and later addition of other species can be done (Kellstedt et al., 2008).
4. **Measurement Sensitivity:** Low-cost sensors frequently measure air qualities with crude accuracy; instability may be acceptable with regard to these measurements if they are designed to indicate poor air quality without precision numerical value (Pleijel, 2009).
5. **Quality Control/Quality Assurance:** The quality of data across monitors can vary greatly. The monitoring plan includes data quality objectives, standard operating procedures, training

protocols, and data management and reporting plans. Monitors also need to be continuously calibrated and retained to correct measuring devices and ensure data quality. The quality of the monitoring data, which may damage public confidence in the data, can be undermined without QA/QC procedures.

6. **Monitor Locations:** Often, the monitoring sites are public areas of local people's interest, such as road intersections, educational facilities, healthcare facilities, public areas, and background sites for understanding regional air pollution.
7. **System of Data Management:** The data monitoring system must archive and track data, such as redundant meta-data recording and meta-data archiving (e.g., location, timing, and quality checks needed to interpret the data). Public data access may also be automated by data management systems.
8. **System Maintenance:** Air quality monitors and data management systems have to be continuously established to ensure that observation and data consistency and accuracy are properly operated without loss of fidelity.
9. **System Costs:** The price ranges from $100 to $200,000 for air quality monitors and additional calibration, maintenance, and operational charges. The total cost of the system will also depend on the monitor density of the network. Guidance on how various types of sensors can be used for various purposes is accessible (Jang & Hart, 2015.

6.2.7. Comprehension of Contributions of Air Pollution Sources

Once monitoring of the air quality has been carried out, monitoring can be used to evaluate which sources of emissions make a contribution most to local air contamination. These estimates of source contributions are crucial to priority mitigation strategies, since major pollutant sources (e.g., power generation, traffic, dust, wildfires, combustion in households and agricultural incinerating) vary from place to place. This type of analysis is often referred to as the "source distribution" because they can estimate or "report" the fractions of the observed air pollution helped contribute by different source types (Hiltner, 2009). Two approaches exist for the analysis of the distribution of sources: bottom-up based on emissions inventories and top-down analysis of source signatures in air monitoring data (see below). The bottom-up method estimates sources of pollution at every location as it is done in a gridded manner, but estimates come from a lack of knowledge on the grid of source emission strengths and simplified description in the chemical transport or dispersion models of atmospheric processes. Top-down outcomes for specific locations, where controls are located, are provided by the top-down approach.

6.2.8. Estimation of Public Health Impacts of Air Pollution

Air pollution quantification of the local public health burden can help to put air quality in context for decision-makers, healthcare professionals and the public. Concentrations of pollutants on their own do not express the health risks of air pollution to the local population. A sufficient granularity allows an assessment of air pollution health impact to estimate the burden of disease and to compare it with other local risk considerations. The input from specific sectors of the emissions source to the local air pollution burden can determine which sectors have the greatest effect on health. An assessment of health impacts can also be used to estimate the smaller number of deaths or illnesses resulting from an action that improves air quality (Figure 6.6).

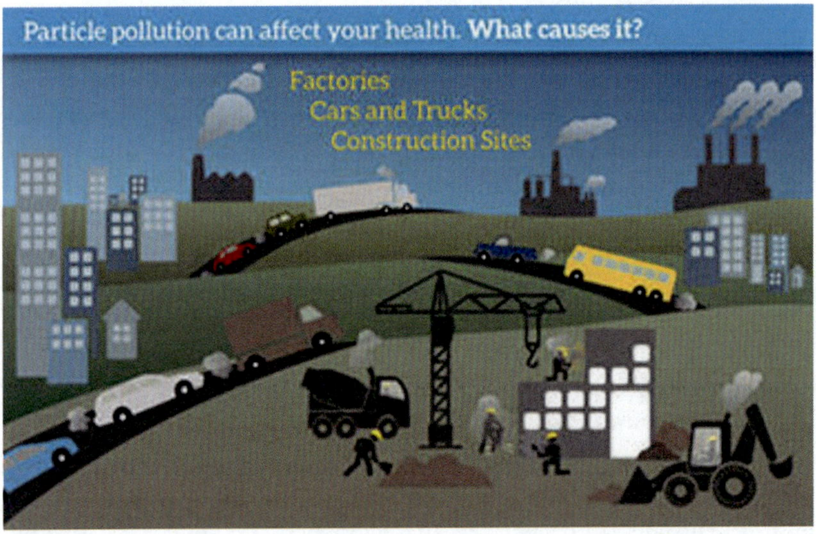

Figure 6.6: Subsequent sources release particle-forming gasses. Sub-source examples include power plants and coal fires. Some other likely causes of particulate pollution may be primary or secondary, such as factories, cars, trucks, and building sites.

Source: https://www.cdc.gov/air/particulate_matter.html.

The following approach is typically employed in health impact assessments:

1. **The Number of Cases of Death or Disease (i.e., Lower Respiratory Infections):** Which are caused by changes in pollutant levels in an exposed population, are contaminant attributable cases of health result.
2. **Changing Pollution Levels:** The difference between the initial pollution load and the emission changeability to concentrate (e.g., implementation of a policy to transition from burning biomass to burning ethanol).
3. **Concentration-Response Factor:** The estimated percentage change in the health outcome risk as a result of the unit change in the level of pollutants generally in epidemiological studies (Jacobson, 2012).
4. **Population Exposed:** Number of people affected by changes in concentration of pollutants.
5. **Incidence of Baseline Disease:** Estimated number of disease cases over a certain duration in a given population (e.g., the number of cases of lower respiratory infection annually per 100,000 people).

A number of tools for automating the health impact analysis are now available. Air quality health impact assessment tools include two overall categories: full-scale tools requiring air quality modeling data for health impact assessment, e.g., BenMAP-CE, AirQ+) and lowered form instruments that connect emission levels to health effects without the need to model chemical transportation; (e.g., LEAP-IBC, TM5-FASST). Generally speaking, reduced-form instruments generate initial results if chemical transport modeling is still unavailable or if many different policy scenarios are compared. Comprehensive models can be used to further evaluate a smaller number of policy scenarios (Schuldt and Roh, 2014).

6.3. WAYS FORWARD

A better awareness of air pollution and health is the next step for the implementation of air pollution solutions. Increased awareness includes measuring and communicating air pollution levels and their health impacts. In this communication, the health sector can play an important role. Other aspects of awareness-raising include the extension and maintenance of evidence, the sharing of knowledge and incentives for enhanced praxis. Air quality management was historically seen as an environmental problem, often not involving the health, energy, transport, and other areas that have influence or effect over air quality. Air quality management has been considered an environment. A new approach to air quality involves the health sector. Although the medical health sector has long recognized air pollution as a risk to health, its direct involvement in air quality planning has historically been limited.

The public health sector has, on the other hand, been historically more involved in identifying air contamination levels and their health impacts through epidemiological studies and developing air quality standards, particularly through research. In sectors, awareness-raising and support for initiatives on air quality improvement in LMICs will have a greater role to play. Key strategies which may be relevant health sector engagement entry points include:

- The investment case for effort to address air quality can be developed by public health professionals by compiling data that link air pollution and health consequences and delimit cost of health care and economic implications. Governments often make decisions on the basis of action costs against inaction or by comparing action costs against action benefits. The Local Health Estimates can provide valuable information for policymakers from different air pollution management actions (for example, utilizing BenMAP-CE, LEAP-IBC, HAPIT or other tools for health impact assessment). Moreover, the live saved tool (LiST) is under discussion in order to make the cost-benefit analysis readily comparable with other public health interventions for these interventions (Hiller, 2011). Hospital data logs, surveys, and insurance data can be used to generate health outcome data, which may not be readily available, to be used in epidemiological studies to quantify the health effect of air pollution. Measuring results at the level of the medical center enables health providers to monitor patients, identify potential community-based risks and educate patients on these risks.
- In messaging and advocacy, medical professionals play a role. Frontline care providers are a strong point of entry for enabling patients with knowledge of atmospheric pollution risk factors and also influencing healthcare by joining other providers in underlining the importance for patients and public health of air

pollution. Appropriate data must be available for the continuing medical education and related programs to ensure that health providers are adequately informed and knowledgeable regarding key risks and interventions. Information can also be disseminated by trade associations and conferences. For example, the Indian Doctors for Clean Air Program offers doctors and medical practitioners the opportunity to speak on air pollution and raise awareness of the effects of air pollution on respiratory and cardiovascular health (Jacobson, 2012). In South Africa, the Pan Africa Thoracic Society trains doctors to learn about the health effects of air pollution and advises patients on their risk reduction. National and local health agencies can also play an important role in this context by developing awareness programs related to broader initiatives in the area of public health surveillance (Figure 6.7).

Figure 6.7: Some areas have low sunlight and are not suitable for solar applications. There is not a lot of wind in others. Regions need to develop what suits them best. The great plains can take wind power as their energy source, while the desert southwest needs to develop solar. Although current nuclear power plants have major security problems and waste disposal, some locations may rely on nuclear power plants.

Source: https://www.dreamstime.com/illustration/air-pollution-solution.html.

- The best practices are advocated by hospitals and medical institutions. Clean sources of energy and clean waste disposal options can be selected by health facilities. If this investment was made into clean energy, it would be both reduced air pollution and a show of the general public that the medical sector consumes many fractions of the total domestic energy. In addition, health facilities at every level (community-to-treatment hospitals), waste generates-both reducing air pollution and providing a model of best practice for the community, from local inciting systems to environment-friendly waste disposal practices. To ensure that health programming includes environmental perspectives, the health ministry can incorporate environmental health mandates. The Healthcare without Harm organization has championed health care systems to choose alternative solutions that are environmentally friendly as health care must not contribute to pollution or disease (Jang and Hart, 2015).
- Inter-ministerial convergence ministries of health can improve planning and implementation of policies through the integration of ministers of health, environment, energy, and transportation. Whilst environment departments can be seen as the natural structure of governments for the management of atmospheric pollution, health ministries can model the planning process; implement policies and programs, host data, and lead structural changes that will eventually connect relevant ministries such as the environment. Ideally, finance ministries with unifying power would lead air pollution programs and policies. Air pollution necessitates the access of information, planning, and response to air pollution through multiple relevant sectors and stakeholders to promote inter-ministerial convergence. The health sector can influence the emitting sectors by emphasizing the repercussions of action versus non-action on public health.

Chapter 7

Effects of Climate on Health

CONTENTS

7.1. INTRODUCTION

Climate is referred to as the average expected weather of a given area over a certain period of time. Numerous variables are used in evaluating expected weather in a given area. They include rain, temperature, humidity, and sunshine among others. It has numerous impacts on both animal and human health. Warm and cold weather have their own effects on health. Climate affects mental, physical, and communication health. Depending on the nature of the climate, extreme weather climates have severe impacts on health. It also increases the incidence of illnesses and diseases. Over the years, their changes in climate have worsened effects on climate on health. It is a cycle in that severe climate affects the environments and in turn it affects human health (Fountain, 2019). The growing increase in global warming is severely affecting human health. There are both direct and indirect impacts of climate on health. Effects are dependent on the type of exposure. When discussing about climate, topics on climate change are unavoidable.

Climate change refers to changes in the expected climate of an area. It makes it hard for meteorologist to accurately predict the type of expected weather. Increase in environmental pollution has resulted in climate change. A major evidence of climate change is global warming. Drastic changes in climate have negatively affected human and animal health, more so mental health. Effects of climate on human health are as follows:

7.2. EFFECTS OF HEAT ON HUMAN HEALTH

High temperatures cause heat stress. There are different sources of heat in the body classified as internal and external sources. Some of the internal sources of heat include metabolic heat. Metabolic heat is produced as a result of body processes. External sources of heat are environmental elements such as sunlight and atmospheric temperature, among others. High environmental temperatures may be as a result of human activities. Some of these activities have resulted in global warming contributing to high atmospheric temperature. The human body is made in such a way that it can accommodate certain amounts of temperatures. However, when these temperatures are exceeded, then it may affect the normal functioning of the body. Temperatures above 35°C make the body uncomfortable exceeding this temperature may result in dire complications (Zhang, 2017).

Normally, the body will try to adapt to the increase in temperature. The body may increase its temperature to try and match that of the environment. This is why the body is said to have an internal thermostat. This ensures all body activities are done as they should. Increasing bodily temperatures requires increased blood flow and sweating rate. There are different ways the human body is exposed to environmental temperatures, namely: convection, radiation, and conduction. Radiation is done when the body is near things that are radiating heat. Convection is when it is round warm air, or it touches warm liquids. Conduction is when it touches hot solid substances.

There are numerous effects of heat on the body. They include high levels of irritation, reduced mental ability to work in different situations and their ability to perform certain tasks. An increase in temperature causes increased blood flow which causes increased heart rates which causes the body to meet important needs. Combination of external and internal temperatures influences the brain's ability to perform certain tasks. High temperatures result in heat-related illnesses. This in turn affects an individual's health. Different factors influence an individual's ability to control heat. For instance, obese individuals may find it hard to control body temperatures. The same case applies to the elderly. Medical implications of heat are dependent on the strength and capability of the body (Haines et al., 2007). Some heat-related diseases include heat rashes. These are small pots usually red in color. They may be located in certain parts of the body or throughout the body. When exposed to heat, the individual may feel as if the rashes are being prickled. These spots are developed as a result of plugging sweat glands. Once plugged they get inflamed resulting in heat rashes. Heat edema is usually exhibited among human beings who are not used to high temperatures. These rashes may be seen on their ankles. The advantage is that these rashes disappear when the individual stays in a cool environment for two to three days. Excessive sweating or loss of salt and water by the body through other means results in heat exhaustion. Some indicators of heat exhaustion include headache, dizziness, breathlessness, excessive sweating and vomiting among others. Treatment requires the individual to stay in a cool environment for a certain period of time. They may be advised to drink plenty of cold fluids.

Heat cramps mostly affect the muscles. The individual may exhibit sharp pain in certain parts of the body having muscles. They can occur singly or alongside other heat-related illnesses. When sweating, the body loses a large amount of salt. While this is ongoing replacement of lost salt may not meet the body's demand. The salt imbalance causes heat cramps. Apart from sweating, heat cramps may be caused by excessive water consumption. People tend to drink plenty of fluids when temperatures are high. After drinking a significant amount of fluids, some of the fluid dilute salt concentration in the body. When salt concentration and consumption is low as this goes on, it causes heat cramps. In some cases, high temperatures cause fainting and dizziness. This is known as heat syncope (Zaval et al., 2014). This is caused as a result of reduced blood supply to the brain while the individual is stationary. It is known to happen to individuals not accustomed to high temperatures. As an individual is sweating, they are losing a lot of water which reduced blood pressure because most of the blood is going to the legs rather than the head. This limits blood supply reaching the brain, making the person dizzy and in some cases, they faint. The best course of action is to allow the individual to rest in a cool environment.

Other heat-related diseases include hyperthermia. This is when body temperatures are beyond normal. It is caused by the body's inability to regulate its temperatures. This is experienced when the body is taking in more heat than the amount it giving out. Hyperthermia is dangerous in that in extreme cases, it could lead to death. Statistics show that about half a million individuals die due to hyperthermia. In severe cases, hyperthermia causes heat stroke. It also affects the body's response to certain medications. Hyperthermia is different from other heart-related diseases in that the body temperature of an individual suffering from hyperthermia is constant. Some symptoms exhibited by individuals suffering from hyperthermia include profuse sweating, breathing complications, dry skin, dilated blood vessels, cardiovascular issues, and nausea (Herndon, 2018).

Treatment includes taking the individual to a cool place and cooling their body temperature using a damp cloth. They should also ensure they drink sufficient amounts of water and refrain from any medication that will affect their metabolism. Individuals most at risk are those less likely to sweat. It is also common among elderly people.

Among some severe heat-related illnesses is heatstroke. There are two main types of heat stroke, namely: classic and exertional heatstroke. Classic heatstroke is the mild form which mostly occurs among vulnerable groups of people such as children and those with underlying health complications. In this case, the individual may or may not sweat. In the exertional case, the individual's body temperature may be high depending on the type of activity they are doing. Symptoms of heatstroke include individuals having temperatures higher than 41°C, profuse sweating, seizure, and loss of consciousness. Some exhibit total or impartial loss of consciousness. This illness requires immediate medical attention failure to which the individual may die. Some illnesses are caused as a result of long-term exposure to high temperature. Illnesses such as damaged kidneys and livers are attributed to long term exposure to heat. It also causes distorted sleep patterns and extreme cases of heat exhaustion. Some research tries to establish a connection between high heat exposure and infertility, more so in men. This is because high temperatures affect sperm production by affecting their density, natural shaping, and motility. Environmental temperature influences the groin temperature above the recommended amount. Research was done on animals to establish the relationship between temperature and pregnancy. It was discovered that high temperatures cause death in embryos. It also causes mutations in during child development resulting in poor brain development. Research is still underway to establish whether the same is exhibited in human beings. However, pregnant women while in their first trimester are advised to stay in areas below 39°C (Singh & Purohit, 2014).

7.3. EFFECTS OF POOR AIR QUALITY IN HUMAN HEALTH

Type of climate experienced greatly affects air quality. For instance, high temperatures negatively affect air quality. This is because warm and high temperatures cause changes in weather patterns. Poor weather leads to air contamination hence poor air quality. An increase in environmental temperatures result in wildfires which creates smoke and releases harmful particles in the environment. Releasing these particles in the environment causes results in illnesses such as asthma. Most respiratory and cardiovascular diseases are as a result of poor air quality (Herndon and Whiteside, 2019). High temperatures and increased release of carbon dioxide in the environment has made most allergens airborne. Certain measures have been put in place to help improve air quality. Changes in climate are making these measures difficult to enforce.

Among the factors lowering air quality includes high levels of ground ozone. Increased amount of ground ozone in the atmosphere is attributed to high temperatures. Ground ozone is a major air pollutant. Individuals exposed to this kind of gas pollutant negatively affects human and animal health. It causes respiratory infections. Once inhaled, ground ozone irritates internal breathing organs resulting in immunity responses such as inflammation of these passageways. It also results in coughing and chest pains. Once it reaches the lungs, it damages the lungs by affecting its normal functioning of the lungs. It may damage lung tissues. In severe cases, it causes chronic bronchitis and asthma leading to hospitalization. Other problems include difficulty in breathing, shortness of breath, and increased incidence of asthma. Once the gas damages lung tissues, it makes the lungs more vulnerable to other infections. Among children, exposure to ground ozone may cause chronic lung diseases. Statistics show that continuous exposure to ground ozone may increase in premature death. It also increases the number of people hospitalized due to respiratory diseases.

Apart from ground ozone, particulate matter (PM) also lowers air quality. They are tiny particles found in the atmosphere examples of PM include smoke and dust. It is generated as a result of the combustion of fossils or fuels. Release of these particles to the atmosphere can be done either directly or indirectly. Indirect emission involves it being generated are a result of chemical reactions in the atmosphere. There are numerous effects of PM in health. Once inhaled, it may cause certain respiratory diseases. Like ground ozone, PM is a major contributor to premature death among children and individuals already having chronicle diseases. Some short-term effects include cardiovascular diseases. Severity of the health complications are dependent on the size of PM. This is because the size determines the level of penetration of these particulate particles into the body.

Coarse particles are able to reach the lungs and damage lung walls. It may enter the bronchi. Fine particles cause severe health effects. These particles are able to penetrate into the bloodstream reaching the brain. Other effects of PM on human health include distortion in heartbeats, increased incidence of asthma, lowered lung functioning and heart attacks. Symptoms of respiratory include shortage of breathe and coughing. Some indirect effects of PM on health include reduced visibility. This is because these particles are involved in the creation of smog. Due to the natural appearance of smog, it lowers visibility. Other illnesses caused by PM include chronic obstructive pulmonary disease. Increased cases of forest fires could worsen the situation. Firefighters are likely to suffer from the negative effects of PM because they are frequently exposed to burning substances.

As mentioned earlier, climate changes cause a release in allergens. Allergens results in asthma attacks. Examples of allergens include dust and pollen, among others. Allergens such as pollen are produced by trees and are transmitted by weather elements like wind. This facilitates mobility of pollen grains. Exposure to pollen causes allergic reactions such as asthma (Hiltner, 2009). It also causes respiratory allergy. When these allergens are inhaled, it interacts with respiratory organs. Once detected by the body, the immune systems respond to it. It initiates an allergic reaction which takes place in two stages. The first stage is referred to as the induction phase. Numerous factors will influence the likelihood of the individual proceeding to the next stage.

Some individuals may not exhibit allergic symptoms. The next exposure stage is known as the effector phase. This is caused by continuous exposure to allergens. Most of these allergens are protein in nature. Once they enter the breathing system, they react with the surfaces causing respiratory inflammation. This is because allergens irritate these surfaces. Warm climate is said to contribute to large amounts of allergens in the atmosphere. For instance, it is in this season that plants ripen and release pollen to the atmosphere. In severe cases allergens cause hypersensitive pneumonitis. This allergic reaction affects the lungs and damages it. Hypersensitive pneumonitis is developed due to continuous exposure to these allergens. Continuous exposure causes sensitivity which later on results in allergic reactions. These reactions damage the lungs as a result of the immunity reaction. Severe cases of exposures may cause chronic inflammation.

Symptoms of hypersensitive pneumonitis are experienced almost immediately as it takes a few hours for the symptoms to be expressed. Major symptoms include breathlessness, fever, chills, and coughs. These symptoms may disappear if the individuals are not exposed to allergens. Total recovery may take a number of weeks. Symptoms such as breathlessness and coughing may take some time before they are exhibited. Continuous exposure to allergens may cause lung fibrosis. Symptoms of lung fibrosis include weight loss, fatigue, presence of sputum when one coughs, and breathlessness during vigorous activities. Extreme cases may result in failure in most respiratory organs. Diagnosing hypersensitive pneumonitis may require a series of tests to be done. Blood tests are done to check for the presence of antibodies in the blood developed as a result of the body's reaction to allergens. Some doctors may require chest x-rays to be done. This is used in determining whether the lung has been damaged. These tests measure the ability of the lungs to carry air while facilitating oxygen intake and removal of carbon dioxide. In cases where results do not give enough information, the doctor may opt to conduct a biopsy.

A biopsy requires one to observe the lungs or studying cells obtained from the lungs. This is done during a bronchoscopy. The lungs may be flushed with some fluids which will carry the cells. Resultant fluid is collected and studied. Treatment may not be required if the case is not severe. However, the individual will be required to keep off any form of contact with allergens. In severe cases, drugs such as prednisone will be given to individuals. These drugs reduce lung swelling. Continuous exposure to allergens may permanently damage the lungs.

7.3.1. Poor Mental Health

Mental health covers the well-being of an individual emotionally, psychologically, and socially. It influences the ability of an individual to feel, adapt, and think. It determines an individual's ability to handle stress. Climate and climate change can contribute to poor mental health. Natural disasters are some of the effects of climate. When experienced, an individual may suffer from trauma due to the loss of a family member or friend. They may also suffer from trauma because of the experiences they had in such type of weather. Events such as hurricanes and drought may negatively have mental health. Direct effects of mental health on the lives of human beings include stress to every extreme. Some individuals takc up comping mechanisms such as consuming alcohol. In extreme cases, the individual may suffer from mental disorders such as anxiety.

Climate may indirectly affect mental health. Weather events like hurricanes result in the destruction of property: homes and businesses included. This results in individuals being jobless or losing homes. This has its effects on mental health. Predictions of poor weather may result in poor mental health. Anxiety will come about due to the fear of the unknown (Hiller, 2011). A high level of anxiety may result in stress. One can therefore state that climate effects on mental health are anxiety-related and in extreme cases, it results in mental disorders. For instance, disasters such as flooding or earthquakes may cause an individual to suffer from anxiety and depression.

Studies have been done to establish the relationship between climate and mental health. For the study, it was discovered that some individuals may develop aggressive disorders due to extreme weather events. Some of them may be violent or abusive to their partners or certain individuals. Climate such as heatwaves have been known to increase the likelihood of an individual indulging in alcohol to deal with stress. It also results in an increased incidence of mental conditions. Certain individuals may possess suicidal tendencies. As stated earlier, climate change may have diverse effects on an individual depending on some factors such as age.

With regards to age, children are most likely to suffer from poor mental health as a result of climate. Some weather events result in the loss of loved ones, property, and jobs. Some children lose their parents to natural disasters. When not given proper care and therapy, they may not recover from the incident. They developmental disorders, and if not treated, such individual may exhibit violent tendencies. Some of these children may be forced to take care of their younger siblings, causing them to have stress at a young age. The elderly, like children, is likely to suffer from mental disorders as a result of climate (Shepardson & Charusombat, 2011). They may suffer from stress or depression due to the loss of a loved one. They may have anxiety due to fear and worry. Pregnant women are also affected by extreme weather conditions. Research shows that women in the last stages of pregnancy are very delicate in that certain conditions may cause them to have early labor. Extreme weather conditions such as earthquakes and hurricanes cause women to deliver early. This is because of the stress and anxiety brought about by these conditions. Some mothers may suffer from post-traumatic disorders. Individuals already suffering from chronic diseases may be severely affected, more so those with mental illnesses. It worsens their mental state. They may also develop other mental disorders.

When individuals developmental conditions, they are more vulnerable to extreme weather. This is because they are given medication that distorts the normal body functions. For instance, it affects their brain's ability to perceive changes in temperature. This influence's the body's metabolism making it difficult to regulate temperature. This may damage the functioning of other body organs, causing them to shut down. In extreme cases, it results in death. Some of these individuals may develop addiction to certain substances. Frequent use of these substances may result in poverty. This makes it difficult for them to cope with these changes. Some depend on services offered by organizations. These services may be cut short due to disastrous weather.

Children are more vulnerable compared to adults. This is because they are likely to exhibit trauma-related symptoms. Children are more delicate in that the slightest changes in routines may negatively affect their mental health. Therefore, separation from parents, moving to other areas, or displacement may cause children to develop stress. However, children are more likely to recover from such incidents over time. Parents or guardians should help their children cope with the effects of climate. They should be continuously evaluated and checked if they exhibit any form of stress (Jang and Hart, 2015).

Among the group of individuals affected by extreme weather events are the first responders. They top the group as they have to play two roles in extreme events; they are responders and victims at the same time. They are tasked with the duty of ensuring aid is availed to victims of such climates while ensuring that they themselves and their family members are safe. This results in anxiety and high levels of stress among them. Others may be harmed in the line of duty, and in severe cases, some may die. This may affect the mental health of immediate family members.

Long-term effects of extreme weather may lead to poor mental health. Extreme climate has caused food insecurity, low cases of employment, and displacement of individuals. Extreme climate such as floods have destroyed homes causing people to migrate. It has also tampered with global food supply causing food shortages. These long-term climate effects have resulted in individuals suffering from poor mental health. Most individuals are stressed and trying to figure out how they can provide for their families. Parents may suffer from chronic stress or depression. Insufficient supply of good quality food to the public may result in poor mental health.

7.4. IMPACTS OF EXTREME WEATHER ON HEALTH

Extreme weather has both direct and indirect impacts on human health. Over the years, there has been an increased frequency in the occurrence of these events. This increased frequency can be attributed to climate change. Events such as floods, severe drought, hurricanes, and storms pose a severe threat to the life of human beings. With the occurrence of these events' certain groups of people, namely elderly people, children, and those with underlying conditions, are at most risk. Among the direct health impacts of extreme events is the creation of mental health problems such as chronic or mild stress, depression, and post-traumatic disorders. Some of these events cause direct harm to an individual's body. People may be hit with objects. For instance, during hurricanes, buildings collapse and fall on people.

Events such as flooding create breeding grounds for insects allowing transmission of diseases. Floods also increase the incidence of individuals more so children suffer from cholera and other water borne diseases. Some indirect health effects of extreme weather include reduced availability of food hence lack of food security. This may cause children to suffer from malnutrition and other food-related diseases. These events also lower water quality making it unsafe for drinking. These events damage infrastructure such as roads and hospitals. This makes it difficult for individuals to access proper health care, which in severe cases will lead to death. They also damage communication systems limiting communication between an individual and health care worker more so those doing first aid. A good illustration is when a child or adult does not know any first aid skills and they cannot access health care workers due to damaged communication systems.

In the long run, extreme weather conditions contribute to increased carbon monoxide concentration in the atmosphere. When these events occur, they cause power outages. To cope with the situation most individuals use electric generators running on petrol or diesel. These generators release significant amounts of carbon monoxide in the atmosphere. This reduces air quality and causes respiratory illnesses such as asthma (Jacobson, 2012). The elderly is very vulnerable during these events. This is because if the evacuations are not properly done, it could result in more health risks. More risks are posed on those with mobility issues. Other challenges include the need for medical records when taking them to hospitals. They also need to be a record of drugs they can and cannot ingest.

Individuals with disabilities, limiting their movement from one area to another, may encounter challenges accessing evacuation routes. This limits their survival chances posing major health risks on them. Those with speaking and hearing impairment may encounter challenges in communicating their needs. They may also face a challenge in understanding warning signs or signals. There are individuals who suffer from shock. They go into a state of shock due to what they have encountered. Pregnant women due to stress may experience early labor causing premature births. Those in their first trimester may miscarry.

7.5. IMPACTS OF CLIMATE ON FOOD SAFETY AND SECURITY

Inadequate food supply and safety directly affects human health. Food is needed by the body to provide energy to do everyday activities. Protective foods enable the body's immunity system to strengthen, allowing it to fight diseases. In children's certain minerals are needed for proper growth and development. Nutrients such as vitamin D helps prevent rickets. Climate has negatively affected food safety. Extreme weather climates have destroyed food supply in globally, bringing about the term food insecurity. Climate has negatively impacted food safety through climate change. It has caused a significant increase in the number of children and families suffering from hunger.

Climate is reduced food supply by affecting agricultural climate. There are certain weather conditions required to grow foodstuffs. With the ever-increasing change in climate, it makes it hard for farmers to identify the type of crop to be planted. Climate change has brought about increased durations of drought. This means that there is limited rainfall supply. Reduced amounts of rainfall means that fewer chances of crop survival. Farmers are forced to take up other measures such as irrigation which is relatively expensive. Certain temperatures make the environment suitable for crop pests and diseases, making them spread through large fields. This reduces crop yields. Among the different types of crops affected are maize and wheat. Statistics show that there has been a decrease in their production (Shapiro & Park, 2015).

Increased environmental temperature is a major factor contributing to the decline in crop yield. Warm temperature increases the incidence of certain diseases. Exceeding hot temperatures as crops germinate result in stunted growth or death of crops. Warm temperatures also facilitate breeding of crop pests. The same applies for cold weather. However, when excessive cold weather is experienced, the implications may be more severe compared to that of warm temperatures. Other weather conditions such as strong winds damage crops. Strong winds can destroy crops and trees by breaking their stems. Wind also facilitates transmission of crop diseases and pests by acting as an agent of transmission. High amounts of rainfall leading to floods also damage crops. Excessive amounts of water may cause the plant to go into a state of shock. It also causes rotting in plants and allows certain pests to breed (Kgabi et al., 2014).

Nutrition mostly revolves around the consumption of vegetables and fruits. Changes in climate is negatively affecting the production of fruits and vegetables. This is evident in the decline in production of fruits and vegetables. This is mainly caused by increased environmental temperature. Heat causes stress in fruit crops causing them to dry out. High temperatures cause plants to develop faster and ripen at a faster rate. This leads to early harvesting, which may result in losses as the harvesting is unexpected. There are certain fruit plants that require exposure to cold conditions. This is making the harvest viable. Continuous high temperatures may affect yields. Cold weather also affects fruit production. Long winters may damage fruit crops and lower their yields.

When discussing food security, animals are also part of the category. Rearing of livestock has also been affected by climate directly and indirectly.

It has been affected indirectly by affecting the production of animal feeds. Poor climate results in reduced production of animal feeds. This results in emaciated animals. In the case of beef cattle kept by pastoralist, effects of poor climate are felt directly. Long durations of drought lead to cattle lacking food. In severe cases, large numbers of livestock die due to lack of food. Warm and cold temperatures also influence the transmission of pests and diseases. Changes in climate also affect milk production in dairy animals. Climate also affects access to food. This refers to the ability of a family or individual to purchase food. Since extreme weather conditions affect food supply, it in turn affects food prices. When there is a limited amount of food supply, farmers, and organizations tend to sell foodstuffs at a higher price. Fluctuation in food prices have been experienced, and some countries have to deal with price shocks. Relatively high prices make it hard for individuals to purchase foodstuffs. This is worsened by the fact that extreme weather causes job losses making individuals unable to supply family needs.

This brings about the relationship between malnutrition and drought. When drought is experienced, individuals may suffer from malnutrition and other food-related diseases. It is one of the direct effects of climate on human health. Drought causes reduced food security which affects the consumption of calories. It affects diversity in food production. This means that the body can be given certain minerals and nutrients depending on the type of food available. Statistics show that a large number of malnourished children are in that state due to increased food insecurity (Kellstedt et al., 2008). Climate has contributed to undernutrition as most of the nutritious foodstuffs are highly expensive compared to processed foodstuffs. Research shows that there are three major ways in which under climate affects undernutrition. They include affecting food security, human health, and provision of health care. It also increases cases of under nutrition due to inadequate food supply.

Extreme weather conditions also limit access to health care services. For instance, children among other groups of individuals are usually given mineral supplements in hospitals. When harsh weather is experienced, it may be quite difficult for them to access these supplements. Climate also limits access to products such as good quality water. Among the pillars of food security is utilization. Climate affects utilization of foodstuffs. Usually, countries tend to control the utilization of foodstuffs to allow it to survive in trying times. When extreme weather is experienced frequently, the country is forced to use some of the stored foods stuff and, in severe cases, may be required to import food from other countries. By tampering with food utilization, citizens may be affected as food supply will be limited, resulting in high cases of malnutrition. Insufficient food supply makes individuals more susceptible to illnesses. They may also suffer from deficiency diseases.

7.6. IMPACTS OF WEATHER ON DISEASES

Diseases affect the normal functioning of the body. Certain weather conditions result in outbreak of diseases. Changes in climate has resulted in increased incidence of disease. Over the years, climate change has been an issue. Research shows that there has been an increase in the spread of infectious diseases due to climate change. This was noted after research was done to establish new disease infections and re-emergence of these diseases. Rainy or warm weather has been said to contribute to the sustenance of most disease vectors. A vector is the mode through which a disease is transmitted. For instance, in malaria, the disease is transmitted from one person to another through a mosquito. The mosquito is the vector. Long seasons of durable weather allows vectors to breed at a fast pace, and they are able to adapt, enabling them to survive harsh weather conditions (Schuldt & Roh, 2014).

Vectors carry disease pathogens from one organism to another. Climate affects vector-borne diseases. The type of climate experienced will determine the presence of these vectors and the spread of diseases. Temperature is a major factor affecting the breeding of these vector-borne diseases. It also affects the working of disease vectors. For most vectors, when temperatures are high, they tend to be active. This means that they can migrate through larger distances spreading diseases to other areas. Low temperatures make vectors in active. Insects such as tick poses this behavioral tendency. When temperatures are they become active, and they move from one host to another. When temperatures are low, they may stick to one host. Exceedingly low temperatures may result in a reduced number of vectors due to death (Kenis and Mathijs, 2012).

For vectors such as mosquitoes, they require moist and warm areas to allow them to breed properly. Rain and sunshine facilitate their breeding. Extreme weather conditions such as floods also create a conducive environment for breeding. This is why people living around flooded areas are likely to suffer from malaria. With regards to diseases, those spread by mosquitoes pose a major threat to human and animal health. Over the years, mosquitoes have been able to pick up various survival mechanisms, allowing them to survive in different environments. Research shows that they are able to survive in the hottest of temperatures. Some mosquito species are able to survive in areas experiencing drought.

An increase in global temperature has resulted in increased numbers of individuals suffering from respiratory diseases. This is because it allows movement and survival of most disease pathogens. Records of individuals suffering from tuberculosis, Hantavirus, and onchocerciasis are high when warm temperatures are experienced. High environmental temperatures cause oceans to be warm. Warm oceans create a suitable environment for algae and bacteria living in the ocean. Some of the algae is toxic in nature. High concentrations of air pollutants such as phosphorous and nitrogen results in a significant amount entering the ocean. High concentrations of phosphorus and nitrogen activate cholera bacteria. This bacteria is known to live in zooplankton. Ocean currents cause water to move allowing cholera bacterial to move from one place to another. This is contributing to cholera outbreak when flooding is experienced.

There are certain diseases whose transmission is heavily linked to climate. Dengue fever is one of them. This is a highly infectious disease. It is also a viral disease. The vector that transmits this disease is malaria. This disease is mostly found among people experiencing tropical climate. With the spread of tropical climate to different areas, the prevalence of this disease is also increasing. Global warming is also a major contributor to the spread of this disease. There are four different types of viruses causing dengue fever. The categories are dependent on the immunity response. For instance, there is a dengue virus that will cause individuals to develop permanent immunity while others develop short-term immunity (Yeager et al., 2011). Headache, skin rash and fever are among the symptoms of this disease. The disease causes dehydration. There is no treatment or vaccine for the disease. Individuals suffering from this disease are advised to drink plenty of water to ensure they are hydrated. Climate change has allowed the spread of the disease to other areas. Weather conditions such as precipitation, warm temperatures, and high humidity allow breeding of mosquitoes. A combination of wet and hot climate fastens the breeding process and allows quick development of mosquitoes. Extreme climates such as El Nino also facilitate the spread of dengue fever.

Spread and prevalence of Leishmaniasis is also dependent on the type of climate experienced. The disease is mostly common in tropical areas hence is a tropical disease. Sandflies transmit the disease. Sandflies usually live on a host and move as the host moves before they move to a new host. There are numerous ways in which the disease manifests itself. In some individuals, they may have ulcers on their skin and as they develop and heal, they form stigmatizing scars. This form of manifestation brings about the name cutaneous leishmaniasis. In the case of visceral leishmaniasis, the parasite enters internal body parts. Leishmaniasis is a vector borne disease hence it is majorly affected by climate. Humidity, precipitation, and temperature influence the spread and transmission of this disease. These conditions affect the breeding and development of sandflies. However, the effects of these weather conditions are dependent on the species of mosquito. There are those affected by an increase and decrease in temperature while there are breeds not affect by temperature changes.

Tick borne diseases also fall in this category. Humidity plays a crucial role in the life cycle of a tick. During the process of growth and development, ticks require high humidity. Temperature can be used to determine tick population. Ticks thrive in high temperatures. It is useful in ensuring proper growth and development in ticks. Climate also influences the likelihood of ticks being exposed to pathogens. As tick develop from nymphs to adults, it is a risk transition because at this stage they are vulnerable to infections. By favoring breeding, growth, and development of vectors climate has its own negative implications on human health. It a factor for disease transmission.

7.7. EFFECTS OF CLIMATE ON OCEANS-HUMAN HEALTH

State of the ocean may directly or indirectly affect human health. This is because a large number of human beings live around the coastal region. Effects of climate on oceans will therefore affect human health. One of the major effects of climate on oceans is ocean acidification. This refers to increased levels of water acidity. Oceans are homes to numerous aquatic animals and plants as it houses numerous species of animals. Climate has also been useful in regulating weather conditions. Oceans have a large area in the world. The vast area of coverage implies that it can absorb numerous amounts of carbon dioxide in the air (Kump, 2011). Release of carbon dioxide to the air by industries and car emissions make this possible. Some carbon dioxide enters water surfaces. Absorption of carbon dioxide by oceans is part of the carbon cycle. It is a method of regulating the carbon dioxide amounts in the atmosphere.

Anthropogenic carbon is absorbed by oceans. Once absorbed, carbon is broken down to form carbonic acid. Carbonic acid acts as a neutralizer in oceans in that it neutralizes alkalinity in water. This results in records of pH of less than seven. Excessive absorption of carbon dioxide causes extreme acidification. By altering water balance, the life of aquatic plants and animals is also altered. Effects of ocean acidification can be directly felt by human beings. This is because they depend on oceans for water, food, and recreation. Eating aquatic food and water obtained from acidified oceans is very dangerous to the health of a human being. Once ingested it may add to the acidity.

7.8. EFFECTS OF CLIMATE CHANGE ON HEALTH POLICIES

Due to the effects of climate on human health, it has created the need for better health policies. Bodies in charge of making these policies are under pressure due to the ever-changing climate patterns. They need to come up with policies that ensure human and animal health is protected. Some of the agencies concerned with health policies have issued reports on the effects of climate change on human health. Over the years, issues of climate and climate change has been a topic of discussion in the global scale because their effects are felt globally. Studies done on climate and its impact on health show that climate affects human health directly or indirectly (Hiller, 2011). A large percentage of people have died due to heat, more so heatstroke, with several others suffering from heat related diseases. Indirect effects include food insecurity causing people to live in hunger. Children are mostly affected, directly or indirectly. This is because compared to adults, children are more likely to suffer from mental disorders due to extreme events, they suffer from malnutrition due to insufficient food supply, and they are more vulnerable to diseases as their immunity is weaker compared to adults.

In conclusion, there are numerous ways in which climate affects human health summarized as follows: climate limits access to health care services. Extreme weather conditions cause power outages and destroy roads. This distorts the normal working of the health care system. It limits the number of individuals able to access health care services. Climate has also caused undernutrition. Climate has contributed to the decline in the amount of calories consumed by individuals. It has tampered with global food supply making health and nutritious food very expensive limiting its access to all individuals (Lineman et al., 2015). Some elements of weather such as heat can be generated by the sun or during human activities. Human beings working in such areas are likely to suffer from health-related diseases. These occupational hazards may limit the individual's working ability. They may carry lesser loads compared to other individuals.

Climate has increased spread of diseases. Warm temperatures have created a conducive environment allowing animal to breed. This increases the population of animal vectors allowing them to spread diseases in other areas. This has brought about the increase in the spread of infectious diseases. Children, persons with disability and the elderly are likely to suffer from mental disorders due to trauma. Extreme weather events such as floods, hurricanes, storms, and drought cause destruction of property and loss of lives (Schellnhuber, 2008). As a result, it causes trauma and grief to other members of the family. Some individuals lose their jobs and lack a means of meeting family needs. This has resulted in many people suffering from mental disorders such as depression and anxiety. Changes in climate have increased the incidence of individuals suffering from respiratory diseases. Warm temperatures caused increased concentration of pollen. Hot temperatures have resulted in increased incidence of forest fires increasing amount of PM in the atmosphere. Availability of pollen and PM in the environment causes increased cases of respiratory diseases as there is continuous exposure to these elements.

Climate has also affected the quality and quantity of water. Floods and strong winds facilitate depositing of dirt and toxic particles, lowering its quality. Warm temperatures make absorption of gases such as nitrogen easier. Large water bodies absorb these waters, causing an increase in levels of acidity. This damages aquatic life, and when ingested by humans, it affects their health. Warm temperatures increase the incidence of individuals suffering from heat-related diseases. Extreme weather conditions like drought have caused death due to starvation.

These numerous effects of climate on human health can be dealt with to a certain level. As most of the effects are due to climate change, relevant stakeholders need to ensure that proper strategies have been put in place to reduce climate change. This includes minimization of human activities causing global warming such as reduced production of carbon dioxide by industries and using alternative means of transport using less petroleum products. The government needs to come up with strategies useful in cushioning citizens after the occurrence of extreme weather conditions. This will help reduce cases of malnutrition, hunger, and mental disorders.

Chapter 8

Solutions to Climate Change

CONTENTS

8.1. INTRODUCTION

The survival of all living organisms depends on the climate. Even the non-living organisms rely on climate for their survival. Climatic conditions do differ from one region to another. This means that the survival of different types of living organisms relies on the different types of climatic conditions; meaning, a certain species dependent on a specific condition cannot survive in a region with different climatic conditions from the one the organism is used to.

The term climate is used to refer to a frequent pattern of weather conditions being experienced in one region. Most individuals confuse the term climate with weather; however, weather is the conditions of the atmosphere in a particular region only for a short duration; whereas climate is the continued occurrence of the same weather conditions in a particular region for long durations (Schuldt and Roh, 2014).

8.1.1. A Brief on Climate Change

With time, the climatic conditions in different regions do change. Factors which contribute to climate change are numerous, ranging from human factors, to natural factors. For ages, the climatic conditions have been known to change, though the change experienced back then was slow change. In our current generation, which is in the 21st century, change in climatic condition is occurring at a faster rate; mainly due to the human activities.

Some of the human factors known to affect climate include activities such as deforestation, use of fossil fuels, use of artificial farm inputs, industrialization, among others. Deforestation is the cutting down of trees in a certain region without replanting them, which leaves the region with the scarcity of trees. Naturally, trees are known to help a lot in the conservation of climatic conditions in a certain region. One, the roots of different trees do hold together soil particles promoting the water retention in the soil. Soil which retains water in the soil help in reducing the rate of evaporation of soil moisture into the atmosphere, ensuring that just the appropriate amount of precipitation is received in a certain region. The effect of increased precipitation due to over-evaporation of soil moisture into the atmosphere results in flooding, which causes catastrophic effects such as destruction of property, loss of life, among other negative effects. Two, trees are known to be windbreakers (Hiltner, 2009). By being windbreakers, it means that they will have to protect different regions from the effects of strong winds such as soil erosion. If trees are cut down, strong winds will have the upper hand, and the rates of soil erosion in different regions will be high. With soil erosion, the soil aggregates will be loosed up, increasing the rate of evaporation of soil moisture into the atmosphere leading to over precipitation hence flooding. Three, trees have been said to be one of the main factors which attract rain. With deforestation, different regions are left with no trees to attract rain leading to drought periods due to low precipitation levels in regions that once experienced high precipitation levels. With low precipitation, seasonal rivers which relied on high amounts of rainfall for the supply of water end up drying up.

The use of fossil fuels such as the likes of petrol, diesel, coal, charcoal, among others, affect the climatic conditions. The combustion of fossil fuels in order to power motor engines or in the provision of energy for activities such as cooking, leads to the production of carbon (IV) oxide among other greenhouse gases (GHGs).

The production of GHGs has been proved to have negative effects on the ozone layer. As the GHGs continue to accumulate in the atmosphere, they rise up to the ozone layer and cause the depletion of the ozone layer (Rothe, 2015). Depletion of the ozone layer is known to cause a rise in temperatures, an effect known as global warming.

The use of artificial farm inputs such as the likes of fertilizers, artificial pesticides, artificial insecticides, and artificial herbicides does contribute to climate change.

This is because, as the artificial farm inputs execute their role, they end up releasing dangerous gases such as the likes of sulfur dioxide (SO_2) among other GHGs. Such GHGs accumulate in the atmosphere, and during the rainy season, the gases react with rainwater leading to the formation of acidic rain. Acidic rain leads to corrosion of house roofs, plant leaves, and acidification of the soil.

Corrosion of plant leaves and other plant parts could cause plant retardation, which causes the same effects as deforestation, for plants are trees, and trees are plants in a way. Acidification of the soil does also contribute to trees retardation, for not all tree species can survive in acidic soils (Rothe, 2015).

Industrialization does also contributes to climate change. Industrialization is the establishment of different industries for the purposes of manufacturing products. During the manufacturing process, some dangerous gases are released into the atmosphere. Most industries have been advised over and over again to treat their emission before released them, but most of them do the opposite of what is expected of them and end up releasing the gases. Gases released from manufacturing industries fall under the category of GHGs, which contribute to the depletion of the ozone layer and also the occurrence of acidic rains.

Natural factors that contribute to climate change are not numerous, but once they occur, most of them take time before their effects are noticed, whereas others are quite catastrophic. One of the natural causes of climate change is volcanic eruptions. Volcanic eruptions cause the destruction of lands by releasing acidic chemical, leaving once productive and fertile regions to be infertile, contributing to the rise in food insecurity crises. They also lead to the production of GHGs as the volcano erupts; which contribute to the rise in temperatures.

The second natural cause of climate change is the change in the sun's energy reaching the earth. If the energy from the sun reaching the earth decreases with time, the temperatures in a certain region will change gradually, leading to a decrease in temperatures; whereas if the energy increases, the temperatures in such regions will increase, leading to change in their climatic conditions gradually.

8.2. SOLUTIONS TO CLIMATE CHANGE

Climate change has caused catastrophic effects across the world. From prolonged drought periods due to low precipitation levels and high temperatures, to increased food insecurity, to skin diseases due to increased emission of unfiltered UV-rays, to flooding, among other negative impacts caused by climate change (Robinson et al., 2009).

With such negative effects, it is high time, the human race focuses on ways of reducing the effects caused by climate change by completely doing away with the causes of climate change, more so the human causes; and do their best to prepare for the natural causes of climate change.

8.2.1. Addressing Carbon Dioxide Emissions

As stated earlier, one of the main causes of climate change is the emission of carbon dioxide gas into the atmosphere. The below solutions address the possible ways in which emission of carbon dioxide, one of the main enemies of climate, can be reduced (Figure 8.1).

Figure 8.1: An image showing the emission of carbon gases from a manufacturing plant.

Source: https://www.google.com/search?q=carbon+dioxide+emissions&client=ms-android-transsion-infinix-rev1&prmd=inv&source=lnms&tbm=isch&sa=X&ved=2ahUKEwiVo9O22OzvAhVCtHEKHVjrCG0Q_AUoAXoECAIQAQ&biw=360&bih=592.

One of the ways of reducing the rates of emission of carbon dioxide gas is by speaking up. The advancement in technology in our current generation has provided us with social media platforms where we are able to interact with different people from different parts of this world. With such platforms, we are able to share our thoughts and ideas on different issues such as the likes of effects of climate change and the most appropriate ways to curb the causes of climate change. Such platforms are a perfect opportunity for climate change activities to voice their concerns on climate change. Individuals are able to provide a space where people can discuss on factors which lead to climate change and be able to unanimously agree on several ways in which they can reduce the rate of carbon dioxide emissions. Individuals in positions of power could also use their position to influence the formulation of laws that limit the rates of carbon emission (Revesz et al., 2014). With such laws, any offenders will be liable to a certain punishment for neglecting the laws, which is an act that encourages other (to be offenders to think twice before emitting carbon dioxide in the atmosphere). In discussions with friends, family members, the topic of reducing the rates of carbon emission should be welcomed. Speaking up creates awareness on effects of emission of carbon gas, the cause of carbon emission and mobilizes people to play a part in ensuring that the rates of emission of carbon dioxide gas are reduced.

Instead of using energy sources fueled by fossil fuels which are the main causes of carbon dioxide emission for with their combustion, carbon gases, among other GHGs are produced, people are advised to opt for the use of renewable energy sources. Homesteads could opt for the use of renewable energy sources such as the use of wind energy which is clean for it does not lead to pollution of the environment during its production; solar energy is so pure and clean for it is basically the energy from the sun which is being converted into energy that can be used to power different appliances at home, in offices and even in industries. If the use of renewable energy sources is a bit challenging, people are advised to find ways in which they can curb the use of fossil fuels to power unnecessary activities. For instance, instead of the use of personal transport every now and then, which consumes more fuel which during combustion in order to power the engine; individuals could use public transport which saves on fuels and reduces the amount of carbon gas released in the atmosphere for the vehicles emitting carbon gas will be few on the road (Hiller, 2011).

Most individuals fill their homes with unnecessary equipment which consume a lot of energy. For instance, in hot regions, people install cooling systems such as the likes of fans in order to cool their homesteads; and in cold regions, individuals install heating systems to warm up their homes and surroundings. Instead of having to install such appliances which consume more energy leading to the emission of carbon dioxide gas in one way or the other, it is advisable that people use energy saving ways to warm up or cool their homes. For instance, in the case of cooling homes in hot regions, people could opt for building spacious rooms for with more space, air circulation will be increased reducing the temperatures; whereas in the case of warming up homes, people could use materials which warm up houses during construction to ensure that there is efficient heat in a room.

Most home appliances have been the main contributors to the increase rate of carbon gas emissions in the atmosphere. However, with the increasing rates of climate change in the world, most manufactures have been forced to reconsider their manufacturing techniques. By reconsidering the manufacturing techniques, improved home appliances have been manufactured, which are more efficient in energy consumption leading to the low emission of carbon dioxide gas. With the availability of such home appliances, individuals are being advised to go for the energy-saving appliances instead of equipment which increases the rate of energy consumption, increasing carbon gas emission. In most nations, energy-efficient home appliances are marked with an "Energy star label," however, the label does differ from one country to another (Pleijel, 2009). Whenever shopping for home appliances such as the likes of refrigerators, washing machines, microwaves, electric gas cookers, among other appliances, individuals are advised to look for the labeled devices which ensure that the rates of carbon gas emissions are reduced.

It is said water is life. However, with such a popular saying being common to everyone, even the illiterate individuals, we, human beings, still find ourselves wasting water. From taking too long in the shower, to leaving water taps running, to using too much water in cleaning up our houses and offices, among other ways. Water wastage leads to an increased rate of carbon gas emission. This is because, from the pumping of water from different sources such as the rivers, water boreholes; the heating up of water for showers; to treatment of water for consumption purposes; energy is used, and with use of energy, it means that carbon gas is emitted. To reduce the rate of carbon gas emission due to the wastage of water, people are advised to develop habits that save on water. For instance, while brushing, one ought to turn off the water tap and only turn it on when using the water; people should train themselves to take short showers which lead to minimal usage of water; water sense appliances have been invented which ensure that there is minimal usage of water when using them and people are advised to go for them.

The other issue which has been contributing to the increased rates of carbon dioxide emission is the increased rates of dumping processed food packages. In our current generation, most individuals prefer packaged foods to homemade foods. With packaged foods, a lot of energy is used in their processing, packaging, and transportation to the distribution stores. The more the energy used, the high the amount of carbon dioxide gas being released into the atmosphere. For instance, during processing, carbon gas is emitted and released into the atmosphere; and also, during transportation, the combustion of fuels to power motor engines leads to the emission of carbon dioxide into the atmosphere (Robinson et al., 2009). As for the packaging, energy is used in the manufacture of packages which does also leads to the emission of carbon gases. After the consumption of packaged foods, the packages are dumped in dumpsites where they pile up and due to the weight of the pile and also the duration they spent on the dumpsite; they end up decaying. The decomposition of packages in a dumpsite does also leads to the production of carbon dioxide into the atmosphere. To reduce the rate of carbon emission due to the consumption of packaged foods, individuals are advised to go for homemade meals which consume less energy in producing them. In this 21st century, most individuals have opted for the modern lighting techniques, which include the use of bulbs. Bulbs require energy for them to light. There are different types of bulbs, some of which consume more energy compared to others. It is therefore advisable that, when considering the use of bulbs for lighting purposes, people are advised to go for the energy-saving bulbs. The light-emitting diode (LED) bulbs are known to be one of the energies saving bulbs compared to the incandescent bulbs, which consume more energy (Jang and Hart, 2015).

Most of the products we purchase for our personal use, home use, office use, among other purpose, do emit carbon gases in one way or the other. With such awareness, manufacturing companies have been advised to always indicate the amount of carbon contents used in the manufacture of a certain product and that they should clearly state the rate of carbon emission from the product with time and under which circumstances. With such information, people are advised to look out for high carbon gases emitting products and avoid them and instead go for products that emit low amounts of carbon gas.

The issue of transportation has been one of the main contributors to increased rates of carbon dioxide emission. The use of trains, airplanes, and other automobiles requires the use of fossil fuels to power their motor engines for movement purposes. As stated earlier, the combustion of fossil fuels in order to provide energy leads to the production of carbon gas which contributes to global warming. To reduce the rate of carbon gas emission, people are advised to opt for transportation ways which do not lead to the production of carbon gas. For instance, instead of having to drive for short distances, people could opt for walking, which improves their body flexibility. When there is a need to use vehicles, people are advised to opt for public transport instead of personal transport, which reduces the number of vehicles on the road leading to a reduced rate of carbon gas emission into the atmosphere. Most people choose flying for short distances due to the speed of planes and the comfort. However, air crafts are known to be the main causes of air pollution; therefore, it is advisable that people opt for normal public vehicle transport instead of flying, reducing the rate of air pollution due to emission of carbon gases.

Most individuals neglect their vehicles. The neglecting of vehicles leads to improper functioning of the vehicles leading to increased consumption of fossil fuels. To reduce the rate of fossil fuel consumption, it is advisable that people maintain their vehicles, trains, and aircrafts and ensure that they are operating efficiently. Vehicles should be services frequently, and if there are any damages on the vehicles, even the smallest damage, the vehicle ought to be repaired immediately. Train rails should be serviced from time to time to ensure that they in perfect condition such that they will not cause any damages to the train (Revesz et al., 2014). Aircrafts should be maintained and taken care of in the most delicate way for any minimal mismanagement of an aircraft could lead to a plane crash which is more catastrophic and does lead to high emission of carbon gas among other GHGs which spike the rate of climate change. Therefore, it is advisable that all vessels of transport, even the water vessels such as the likes of ships, be serviced frequently and maintained in perfect condition to ensure that they do not consume more fuel leading to high rates of carbon gas emission.

Still under transportation vessels, there exist different models. Some models are known to consume less fuel, whereas other models are known to consume more fuel. It is advisable that people be keen on the rate of fuel consumption before the purchase of cars. Same case does apply to trains and aircrafts. Each model has its own advantage and disadvantage, but the main factor to be considered in the rate of fuel consumption; for the more the rate of fuel consumption, the higher the rate of carbon gas emission.

We are in a new technological era, where almost every appliance requires electricity for it to operate. For this reason, most of the time our home power outlets and office outlets are always plugged in with different appliances. The habit of plugging appliances into electricity has gotten into us to the extent that some individuals forget to unplug fully charged devices. For this reason, it is advisable that individuals develop the habit of always unplugging the fully charged devices and switching off the outlets which are not in use (Lineman et al., 2015). This way, electric energy will be saved, leading to slow emission of carbon gases into our surroundings. Still, in order to save on electric energy, individuals are advised to adjust their devices to consume less energy when not in use, and if possible, adjust their devices to still use the least amount of energy when in use. This way, appliances will take time before they need a recharge saving on consumption of electric energy.

8.2.2. Proper Land Management

The issue of land use has been overlooked, yet in so many ways, it does contribute to climate change. For instance, land is used for agricultural purposes, and most of the agricultural activities being carried out in different farms have contributed immensely to the increased rate of climate change. Trees are also grown on land, and deforestation, which is then cut down of trees as stated earlier, has contributed to climate change in so many ways. Therefore, it is high time people agree on the fact that lands need to be properly managed to curb the rate of climate change. Some of the ways which could help in proper management of land include the conservation of lands and all that the lands hold such as the likes of trees; restoration of destroyed lands through deforestation or uncoordinated mining process leaving behind open ditches on the ground; and the improvement of the already existing strategies used in land maintenance.

On the issue of conservation, forest ought to be on the number list of conserved land resources. Natural forests are known to be the main harbors of carbon gases. Natural forests find their own ways of balancing the amount of carbon gases in the atmosphere and ensuring that only the required amount, is available in the atmosphere, for it is the excess carbon gases that lead to climate change. Natural forests contain so many trees which use the excess carbon gases for photosynthesis purposes which is the manufacture of plant foods. If the trees are destroyed, the excess carbon gases will have no use and instead they will continue accumulating in the atmosphere leading to the destruction of the ozone layer which leads to rise in temperatures. Most natural forests are known to be the sources of different rives, both permanent rivers and seasonal rivers (Kump, 2011). The destruction of natural forests means that most natural rivers will end up drying up, changing the cool climates in different regions due to the presence of such rivers. It is therefore advisable that natural forests be conserved. If possible, natural forests should be set aside as national reserved restricting any encroachment activities by locals residing close to natural forests. With most natural forests being sources to different rivers, it means that they control the flow or water all over; hence helping in mitigating the occurrence of frequent floods in different regions. So, the conservation of natural forests is of great importance for helps reduce the rates of climate change saving the world from a lot of agony.

Restoration of land resources does help in curbing climate change. One, with reforestation, deforested areas will have the change of producing life increasing the soil productivity. Reforestation restores the beauty of a once ugly bare land. As stated earlier, trees do help in regulating the amount of carbon gases in the atmosphere; this means that, with reforestation, regions which had an imbalance in carbon gas levels will have the chance of holding just the required percentage of carbon gases which is not harmful in their atmosphere. This is because trees will be available in order to use the excess amount of carbon gas in the manufacture of their plant food (Pleijel, 2009). Still under restoration of land, grounds which have been left with lots of open pits after mining practices should be filled. This is because most of the open pits tend to be way too close to the lower layers of the earth which emit poisonous gases in to the atmosphere leading to climate change. The filling up of such pits means that the surrounding will not be exposed to such underground GHGs. Basically, the restoration of wasted lands, be it through reforestation or filling up of open pits, among other practices, does help a lot in lowering the rates of climate change.

When it comes to proper management of land, people have to focus more on the agricultural activities for they are the main activities which require land to be maintained properly. Most of the agricultural activities being conducted in our current generation instead of maintaining land, lead to its destruction. For instance, the use of artificial fertilizers leads to degradation of the soil quality by lowering its fertility rate and the destruction of most of the soil microorganisms, which help in improving soil quality. Also, the land preparation techniques, instead of preserving the land, they end up destroying it. Therefore, it is advisable that most of the farm owners do their best to ensure that their agricultural activities align with proper land management techniques. For instance, instead of the use of artificial fertilizers, which affect the health of the soil negatively and lead to emission of GHGs; farmers are advised to opt for compost manure which is natural, and instead of lowering the soil quality, it boosts it. When it comes to land preparation techniques, farmers are being advised to go for the direct methods of plantation. With direct plantation, only the part of the ground where the seed is to be planted is disrupted. With minimal disruption of the soil, the soil structure is maintained hence the maintenance of the health of the soil (Kenis and Mathijs, 2012).

To help in the restoration of deforested areas, governments should make it their top priority to fund projects which promote reforestation. If possible, governments officials ought to be in the front line leading with examples by planning tree planting days frequently in different parts of their nations. This way, the rest of the country people will be encouraged to plant trees every now and then, all over their surroundings.

Natural resources such as the likes of rivers and lakes are known to have immense capabilities of storing excess carbon dioxide gas in the atmosphere. Such natural resources should be restored and allowed to flourish naturally.

8.2.3. Addressing the Use of Fossil Fuels

As it has been stated earlier several times, the use of fossil fuels is known to be one of the main causes of climate change. This is because, for one to use fossil fuels, in one way or the other they will end up being burned in order to produce energy; and the burning of fossil fuels is known to lead to the production of GHGs such as the likes of carbon dioxide, SO_2 among other gases which contribute to climate change (Figure 8.2).

Figure 8.2: An image showing the burning of coal.

Source: https://www.google.com/search?q=fossil+fuels&tbm=isch&chips=q:-fossil+fuels,g_1:coal:w39G3gaDsRc%3D&client=ms-android-transsion-infinix-rev1&prmd=inv&hl=en&sa=X&ved=2ahUKEwiF6Z-Z2ezvAhUV4oUKHTIADU4Q4lYoAHoECAEQBA&biw=360&bih=648#imgrc=A45rlRp-2Iar4M.

One of the ways to address the contribution of fossil fuels in climate change is by burning the combustion of fossil fuels. The burning of coal and charcoal should be completely burned for with their burning, there is an increase in the amount of GHGs leading to temperature rise, acidic rains, drought periods which in turn leads to food insecurity. If possible, nuclear energy among other renewable sources of energy such as the likes of wind energy, solar energy and geothermal energy should be used instead of coal; for such energy sources are known to be secure and free from any form of waste which could contribute to climate change (Hiltner, 2009).

The burning of oils should also be prohibited. Oil is present in several products used in our day-to-day life. For instance, all plastic materials contain some percentage of oil; from the plastic packaging tins to the polythene papers mostly referred to as nylons, among other products. Campaigns should be held frequently to help in educating the public on appropriate ways of disposing of plastic waste for burning of plastic waste pollutes the environment lowering the soil quality and changing the climatic conditions in a certain region due to production of poisonous gases which might lead to retardation of both plant and animal life.

The use of fossil fuels to power cars should be addressed. Electric cars should be embraced in the society for they do not lead to the release of GHGs; reducing the rate of climate change caused by increased percentage of GHGs in our surroundings.

With the advancement of our current technology, it is advisable that individuals invest their time and energy in conducting research activities which will lead to the invention of better energy sources to replace the use of fossil fuels. For instance, the use of biogas as a source of energy should be perfected such that in the future, it will be the cheapest and easily available source of energy. The already existing solar energy and wind energy should be advanced such that in future, there power levels will not be solely dependent on amount of sunlight or speed of winds in a certain day, respectively (Hiller, 2011). The use of nuclear energy should be perfected, for currently, any mismanagement of nuclear power plants can lead to catastrophic effects which could last for centuries. Basically, to fully address the issue of climate change, a lot of time, energy, and resources ought to be invested in the invention of better sources of safe and clean energy.

8.3. ADDRESSING CLIMATE CHANGE IN OUR HOMES

Most of the products used in our day-to-day lives are known to contribute to climate change. For instance, at home, people are being advised to purchase only the required amount of home shopping to avoid the dumping of expired products because they were more than enough; and as stated earlier, the dumping of products, more so the processed products do lead to decomposition which contributes to climate change due to carbon emissions from the dumpsite.

As home and also in the offices, people are advised to cut on the usage of electricity by; switching off the lights when not in use, unplugging the fully charged home appliances from home outlets, among other practices.

To help in the preservation of our environments, people could be advised to use reusable materials for different purposes. For instance, instead of using serviettes in wiping tables, people are advised to use table cloth which can be washed and reused again. This is because, serviettes used to clean tables will end up being thrown away increasing the amount of waste materials in our surroundings, and the more the waste, the more the increase in production of GHGs from decomposing waste, and the increase in the rate at which climate change is occurring (Kenis & Mathijs, 2012).

When it comes to shopping, most individuals find themselves purchasing small items day in day out, which costs them more; and in terms of climate change, increases the amount of packaging bags to be used which are known to contribute in one way or the other to climate change. Therefore, it is advisable that people develop the culture of doing their shopping in bulk. With bulk shopping, only a few packaging bags will be used hence reducing the number of them to be thrown away; which is a practice that saves on production of GHGs from decomposing packaging bags.

8.4. REDUCING WORLD POPULATION

This might sound weird, but it is a true fact that, the more the people living in this world, the more the rates of pollution, which are the main causes of climate change. The issue on rising population is not that they will be a burden or will deliberately cause the pollution stated above; but the fact is, most of the families which do not believe in family planning, find it hard to survive because most of them struggle in providing the most basic needs. This means that most of them will find themselves destroying the environment as they do their best to find easy ways to survive (Jang and Hart, 2015).

One, natural forests will be cleared in order to accumulate the growing population; and stated earlier, the clearance of natural forests, does lead to an increased rate of climate change for their will not be trees which will help in maintaining the balance of carbon gas in the atmosphere. Two, the rate at which the environment will be polluted will increase, for with more people, there will be an increase in waste materials. An increase in waste materials means that waste will done almost everywhere, which increases the rate of climate change; for as the waste decomposes, there is an increase in GHGs.

The issue on population does not mean that there is a specific number of people required to live on earth. But the fact is, family planning should be encouraged; for with family planning, children will be able to get the proper attention from their parents and teachers; hence they will be more disciplined and goal oriented. Children brought up in a disciplined environment will have the chance of embracing the environment and doing their best in ensuring that they conserve the environment hence reduced rates of climate change.

8.5. ADDRESSING NATURAL CAUSES OF CLIMATE CHANGE

The natural causes of climate change are way beyond human control. However, with the research being conducted in different parts of the world, there is some light at the end of the tunnels, which promises to help in finding ways to solve climate change caused by natural factors. A technique known as geoengineering is used.

Geoengineering is basically the process of use of human techniques to reverse the effects caused by natural factors. For instance, if the amount of solar energy from the sun reaching the earth increases, chemical particles such as the likes of sulfate particles could be introduced into the atmosphere. Sulfate particles are known to have a cooling effects in nature. The cooling effect helps in reducing the amount of solar energy reaching the earth by cooling the sun rays up in the atmosphere. Apart from the use of sulfate particles, mirrors could help in diverting the effects of increased solar energy by the sun.

Mirrors could be sent to space and placed at strategic position, based on research from astronauts, to deflect the sunlight from the sun. The deflection of sunlight from the sun prevents the excessive sun rays from reaching the earth's surface, reducing the amount of solar energy reaching the earth (Kenis and Mathijs, 2012). Without the current level of technology as the human race of the 21st century, it is possible to cover some of the portions of our planet earth with reflective materials which will help in reflecting back of excess sun rays from the sun reducing the amount of solar energy reaching the earth hence eliminating the chances of climate change. Scientists have managed to invent chemicals which contain iron and another nutrient that could be used to fertilize our oceans and seas. Fertilized oceans increase the rate of evaporation of seawater which leads to the formation of cloud cover increasing precipitation levels which help in cooling down the temperatures caused by increased solar energy from the sun. Still, with increased cloud cover, the clouds are able to reflect the solar energy from the sun. The chemicals used in the fertilization of oceans are known to have the special effects of harnessing excess carbon gases and storing them in the deep-sea beds.

8.6. ADDRESSING THE INCREASED RATES OF FREEZING

Due to climate change, which has led to the rise in temperature levels, the rate at which glacier at the top of the high mountains, the Arctic and Antarctica has been melting has increased.

To help in reducing the rate of the melting of ice, more so in the Arctic and Antarctica regions, the use of giant water cannons, as proposed by the United States scientist would help a lot. This is because with an increased rate of melting ice, the seawater levels rise, which affect the climate in different regions in one way or the other.

Giant water cannons, according to the US scientist, would be used to spread seawater onto the surface of the ice layers. The seawater would then freeze on top of the frozen ice layers, thickening the ice layers. Thick ice layers help in the reflection of excessive sun rays from the sun before reaching the earth's surface, reducing the chances of rise in temperatures which lead to climate change gradually.

The issue of the use of giant water cannons to help in preventing rise in temperatures and increased climate change is still under experimentation, but its perfection with time could be a great boost in solving climate change (Kump, 2011).

To help in addressing the issue of volcano eruptions which are one of the main natural causes of climate change, governments are advised to invest in research activities that focus on ensuring that volcanic eruptions across their country is week understood. For instance, with enough research and experimentation practices, the active and the dormant regions could be easily identified early in advance. The government will be able to detect the chances of occurrence of volcanic eruptions and find ways to prepare for it. If possible, with continued research, ways could be invented to ensure that the effects caused by volcanic eruptions are controlled to ensure that they do not contribute to climate change.

In conclusion, climate change is one of the main catastrophes facing the entire globe. Thorough research should be done on the possible causes of climate change, ways in which the causes of climate change should be solved; for a brighter future of the entire world, for it depends on it.

Chapter 9

Adaption to the Impacts of Climate Change

CONTENTS

9.1. INTRODUCTION

The climate is changing already: since pre-industry times the average global terrestrial and ocean surface temperature has continued to increase by 0.85° C, while ocean acidification is increasing by 26%. Certain forms of severe weather events are increasing and chronic (IPCC, 2014). These changes exacerbate the risk level for companies and economies, such as food production pressures attributable to adjustments in farm yields and the potential to flood cities due to more extreme precipitation (Mika et al., 2018). Climate change may also result in new risks, like the spread of vector-borne pathogens in originally free areas. In addition, the higher average global temperatures, the higher the chance of disastrous changes. The melting of the Himalayan icecap glaciers, the collapse of the ice sheet of Greenland, and the death of the rainforest in the Amazon can include these prevalent and irreversible effects.

The possible impacts of climate change are illustrative of the intense weather events. Economic losses resulting from meteorological events (e.g., floods), climate events (e.g., heat waves) and hydrological events (e.g., floods), have been rising. Climate change, however, is expected to become a more critical component for future losses largely due to the growing exposure of people and economic assets to climate and weather risks. These loss forecasts only constitute part of all-weather event costs, as the decline of ecosystem services and long-term effects on health do nothing to involve non-market impacts. Climate change has already shown its impacts throughout the United States. Some of the severe weather events like heat waves are more common and severe, cold extremes are less common, and precipitation patterns are likely to change. For instance, throughout the western United States, the proportion of precipitation that falls as rain instead of snow has doubled. Over the last 30 years, arctic sea ice has decreased considerably (Figure 9.1).

Even if greenhouse gas (GHG) emissions were reduced significantly now, the climate would remain changing for a long time, and the possible repercussions would be significant for human existence and the ecosystems. For instance, it is approximated that the 2003 European heatwave, if occurred now, could cause an approximate of five times the mean number of deaths related to heat in comparison to the previous time, in a city such as New York, a large American city, for just this one event.

Figure 9.1: Among the most complicated problems that we face today is climate change. It covers numerous dimensions—science, economy, social system, politics, ethical, and moral matters—and is a global issue that will be around for years and decades to arrive, felt at local levels.

Source: https://climate.nasa.gov/solutions/adaptation-mitigation/.

With one such event on its own, the approximate number of deaths in New York City could surpass the existing average national summer. In ecosystems, changing climatic conditions could modify plant and animal distribution trends, reduce species efficiency and ample supply, and alter habitats (Milfont, 2012). The sea level has risen more rapidly than ever, and the natural and man-made environments on the coastlines and in natural waters have threatened, particularly in combination with the impacts of more severe coastal storms.

Adaptation refers to actions that could be taken by nations, societies, and companies to reduce the threats posed by climate change. In the longer run, mitigation-a dramatic reduction in GHG emissions-can only fight climate change, and the faster it occurs the minimal adaptation it will take. Even if all emissions of GHGs halt today, the planet would remain warm, because it has been hundreds of years of GHGs in the atmosphere. This makes it even more crucial today to adapt. Adaptation measures cover large-scale infrastructure projects and community-based small-scale reforms. Research and activities, data, and alterations in livelihoods, actions, and business practices are included. According to the UN Environment Program, adaptation costs in Africa by 2070 are estimated to reach US$ 360 billion every year (Kellstedt et al., 2008).

For example, it will be crucial to adapt to the severe and less predictable climate, which can significantly raise food production and reduce crop failure. Including:

- Changes in crops and animal species tolerating more variable and extreme weather conditions. A different variety of maize has started to be utilized in Zimbabwe that grows faster than conventional varieties, needs lesser water, and produces two dough rather than a single dough. The other has an opportunity to develop if one cob fails.
- Diversifying of livelihoods to include a wider array of plants or farming mixes.
- Enhanced climate prediction systems and extreme warnings for farmers. Farmers in Ghana install rain measurements to record monthly precipitation and better determine long-term trends. Niger aims to update its observatory on climate change and advanced warning systems and make it possible for growers and others to gain better climate data.
- Modifications in the way food is stored and transported. In parts of Malawi, the people of heroism have moved from the storage of grain in granaries to the use of smaller harvest bags, which are easily transferred to higher ground if there is a flood (Peter, 2018).
- Changes in water supply management, which include rainwater collection and storage.
- Index-linked insurance plans compensating growers and herdsmen for climatic events, such as a predefined number of consecutive rain-free days.
- Ducks are raised in floods better than chickens, and elephant grass is planted to slow water from floods along the shore of the river.
- To resist extreme weather and coastal inundation, Mozambique intends to consolidate rural roads and reintegrate irrigation and drainage structures.

9.2. CHOICES OF ADAPTATION

The need of society to face weather extremes and altering environmental circumstances is not new; since the start of time, people have adapted to their environment. One of the oldest instances is agriculture, as farmers had to modify their growing and breeding practices constantly to different climatic conditions. The advancement of flood plain regulations, insurance policies, wildlife reserves, beverage reserves and edifice codes are part of modern attempts to sustain and safeguard everyone's home, livelihoods, and assets in the face of varying climate change. These measures have, even so, been taken in response to the stable climate in the last 10,000 years (Jang & Hart, 2015). The preparation of adaptation to climate changes (instead of climate variability) is now underway in many sectors which are likely to be influenced, from agriculture to tourism. Philadelphia, for instance, has intended an early warning system for hot waves, especially for the vulnerable groups of the elderly and the poor, to reduce related diseases and deaths (Figure 9.2).

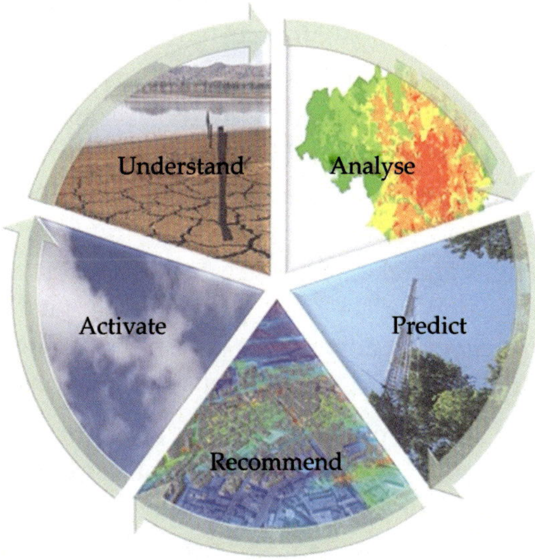

Figure 9.2: Our basic science research must be implemented and adaptation systems to manage existing global social and physical situations. This involves the study of atmospheric compositions, encounters between rainforest and environment, urban reactions to temperature and sea levels rising, as well as natural hazard effects.

Source: https://www2.le.ac.uk/colleges/scieng/research/centers/ace.

In the short term, the most easily deployed adaptive measures are low-cost methods, which have win-win results, short-term co-benefits, and measures to end or reverse poorly adapted practices and policies. At national level, short-term opportunities involve possibilities to modify existing programs that address future climate change predicted; instances include nationwide flood insurance programs; guidelines of federal, state, and skilled engineering; and the management system in the coastal area (Philander, 2008).

The study outlines various potential adaptation strategies-many of which are extensions to address climatic changes-to be considered for short-term intervention in key areas of ecosystems, forestry, and agriculture. Regrettably, many options lack sound advantage, cost, potential, and restrict information for three purposes:

- Being unable to clearly attribute to climate change's numerous expected changes at regional and local level (and thus to record how adaptation causes effects in lowering the impacts);
- The variety of effects and vulnerabilities in the US;
- The comparatively small research group focused on actions to adapt to climate change.

Decisions made so far to deal with climatic variability will be of limited value if attempts to limit pollution fail to deal with the effects of large and rapid climatic change projected. Sudden shifts that transcend the global climate are feasible and create new, possibly irreversible circumstances, like ice-free Arctic summers and severe rising sea levels. Adapting to these effects could entail massive changes in society and government and deliberation of initiatives of adaptation, currently inappropriate or at least extremely difficult, like the mass movement of populations from risky areas.

These decisions and the problem related to funding adaptation attempts are already facing Alaskans. If there are sufficient impacts to prevent a return to pre-disturbance conditions, some existing guidelines, and regulations, like ecosystem and land management policy, may need to be reconsidered and revamped. Cautious risk management indicates that contingency plans for low-probability events, endorsed by further research on the implementation of "worst-case" plans, need to be considered. Adapting to climate change requires a new paradigm that handles climate change threats by identifying prospects for variations, trends, and historical conditions. This implies that we do not wait for inconsistencies to be decreased to include adaptation measures. Mobilizing to boost the adaptive capacity of the global community can now be a policy against an unknown future (Jacobson, 2012).

Climate change susceptibility is often influenced by the following factors:

- The aspect and extent of the alterations experienced.
- Cultural, societal, financial, geographical, and ecological factors underlying climate change sensitivity.
- The capacity of the global community to prevent, prepare, and respond to environmental, financial, and human systems effects.

The first phase for decision-makers is to identify key weaknesses for a region or industry. The next step is the identification of several adaptive options. The range of adaptation strategies that will finally be considered worthy will vary significantly from one region to another and from one sector to another. Decision-makers must weigh the relative costs of adjustment options against the affect risks and identify which possibilities will contribute to other management objectives, like sustainably developing, promote community health, and increasing economic competitiveness. Decision-makers will always need to evaluate the impacts and keep updating their adaptation decisions as new information about climate change, weaknesses, and adaptation efficiency arises (Pleijel, 2009).

Certain civil and resource managers have been involved in adaptation planning by now. New York City accepted such a procedure as a means of integrating continuing plans focusing on development control, infrastructure, and sustainable environment. As an integral element, climate change was selected as a focal point for adapting to climate change risk. This "continuing integration" of climate change plans into other current initiatives has quickly moved adjustment towards an advanced stage. Capacity building involves not only infrastructure building and regulations, but also much more adaptable organizations and investment opportunities in adaptation research and results research. The study highlights that the federal government pursue a substantive future research on climate change adaptation to explore new actions and significantly enhance understanding of the advantages, costs, prospective, and possible limits of such option.

9.3. DIMENSIONS OF CLIMATE ADAPTATION

Both public and private players take adaptation strategies through initiatives, technology, and infrastructure investment and behavioral changes. Governmental action, for instance, will be needed if the government is to operate as a property owner (roads, bridges, etc.), or if major social institution is to be found in need of such action (flood control obstacles or irrigation platforms (e.g., adjustment to regulatory framework, information provision)). But if private citizens act independently, adjustment is regarded rather private to reduce the economic loss from climate change (Hiltner, 2009). The time, the purpose and the real motivation of its execution is another way of differentiating between adaptation forms. Anticipatory adaptation relates to activities taken in ahead of time of effects that are measurable, whereas reactive adaptation is implemented after original climate change effects have been observed. In biological ecosystems, adaptation is naturally reactive, but both anticipatory and reactive in social beings (Figure 9.3).

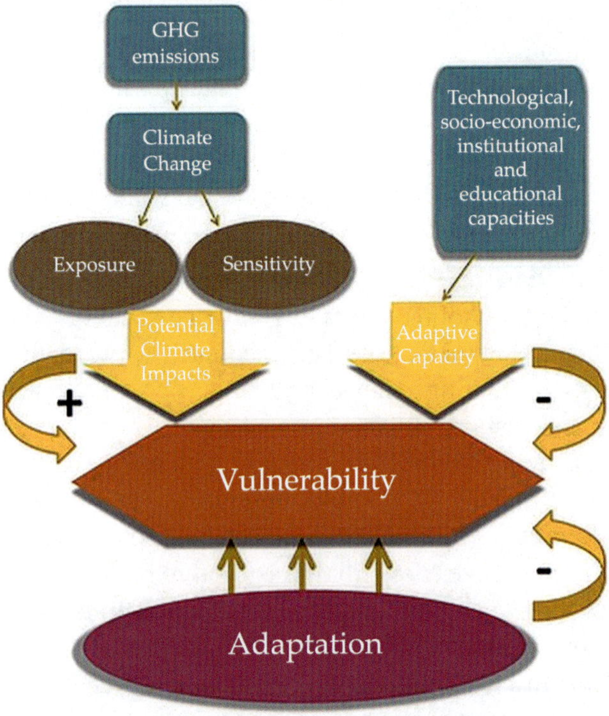

Figure 9.3: The climate change method creates vulnerability and adaptation that attempts to decrease vulnerability. Climate change influences both exposure and responsiveness, which in turn increases vulnerability and thus generates prospective environmental impact. The possibilities for vulnerability reduction are response and adaptive capacity.

Source: https://climatepolicyinfohub.eu/climate-change-adaptation-needs-barriers-and-limits.

9.4. PLANNING FOR ADAPTATION

Future uncertainty must not be an obstacle to climate change preparedness. There is a better understanding of many elements of the climate (e.g., increasing temperatures) than others (e.g., precipitation shifts), but they are all susceptible to certain uncertainties (IPCC, 2014). Climate risks also result from complex and quite often uncontrollable climate- and environmental systems encounters.

For instance, it would be caused by various factors, such as urban design, building construction, and the efficacy of emergency planning whether extended high temperatures contribute to excess deaths. With the attributes of risks progressively difficult to anticipate over time, a proportionate, adaptable, and iterative method for managing risk in the adaptation planning should be used to provide policy responses. Including:

- Improvement by national assessments of understanding about the risks of climate change.
- That use these assessments, several possible outcomes are planned and not a 'most probable' estimate.
- It is unrealistic to recognize zero risk and to prepare the recovery and response systems for the remaining risk.

Inherent links between the climate risks and the actions taken to resolve them are other policy measures. By incorporating them into relevant policy procedures and legislation cycles, for example, with regards to land use management and planning of resources, the effectiveness and efficiency of adaptation planning can be enhanced. In identifying that adaptation is not inherently a dominant policy objective, mainstreaming helps ensure that adaptation goals are associated with policy objectives. This prevents certain potential climate adaptation misalignments, for example, infrastructure regulatory regimes that prevent investment in adaptability; policy planning to promote advancement in vulnerable areas; as well as natural resources underpricing (OECD, 2015c).

Climate change preparedness has become progressively national and strategic in countries. Actively, 24 OECD countries have reported, and 7 are in the process of creating national adaptation measures. Under the UNFCCC National Adaptation Plan (NAP) process, 52 developing nations, mainly LDCs, have presented or are implementing mid-to-long-term adaptation strategies (UNFCCC, 2015). At the same time, local, and community planning, driven by organizations like ICLEI, C40 and World Mayor-

Council on Climate Change, has been increasingly important (Revesz et al., 2014). Policymakers must acquire and gain knowledge from each other, at all stages of development, to maximize the assets invested in adaptation.

This is reassuring, but advancement is still possible. The adaptation strategies of the OECD countries are all premised on impact studies, almost always based on historical developments and climatic situations, and many have classified adaptive capacity, but few have taken adaptation to the plans or systemic responses established. Only two NAPs were presented. A thorough risk and vulnerability evaluation has been undertaken by the UNFCCC (Figure 9.4).

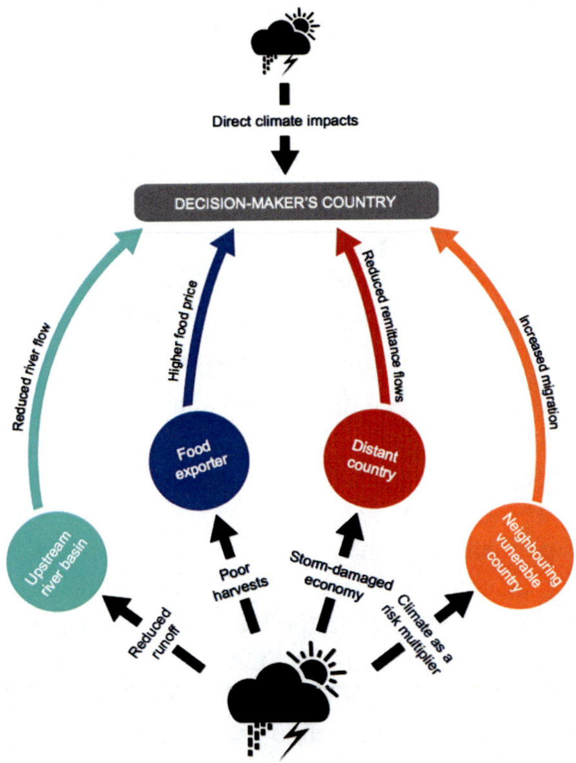

Figure 9.4: To date, just several nations have deliberately considered indirect climatic changes, conversely choosing to focus on their territories' immediate effects of climate change. Our emphasis here is on concrete actions planners can take to identify key indirect effects and to start integrating action in their NAPs.

Source: https://www.weadapt.org/knowledge-base/transnational-climate-impacts/national-adaptation-plans-and-the-indirect-impacts-of-climate-change.

9.5. NEEDS FOR A STRATEGY ON NATIONAL ADAPTATION

As climate change weaknesses and adaptive capacity are disparate and often local in nature, adjustment choices need to be created and enforced by many stakeholders at all levels of government, enterprise, and society. Although contexts are different, decision-makers often understand from their peers' perceptions and evolve best procedures to their own situations. Even so, there is officially no reliable federal policy directive promoting proactive climate change adaptation or a place for supervisors to express innovative ideas and work together for adaptive work.

The federal government must perform an important role as driving force and organizer in overall planning of this void: giving information, necessary expertise, and encouragement for decision-making and application of adaptation; avoiding unnecessary unintended consequences and inconsequential or ineffective investments and outcomes. Therefore, the review suggests that the executive branch create a national adaptation strategy in deliberation with Congress. This method would be enacted by means of programs aimed at guiding adaptation across jurisdictions, providing support and guidance to the scientific research required to broaden understanding of effect and adaptation (Robinson et al., 2009).

9.6. IMPLEMENTATION OF ADAPTATION AT THE SECTORAL LEVEL

The threats associated with water and water have a major impact on economic growth. Most country efforts have so far focused on developing evidence-based evidence and trying to develop information-based tools like flood hazard maps as well as adapting guidelines to local governments to improve their water security. There are also countries that revise regulations and laws such as the boundaries on sustainable water extrapolation, construction codes and land use plans. They also adapt financial instruments to decrease baseline stresses, raise funding and resolve the growing risks of floods like water tariffs, water-related environmental taxes, and flood insurance systems (OECD, 2013).

Change in climate is going to affect many other agricultural commodities' productivity and costs. The OECD has discovered that independent adaptation could not be adequate to prevent all damage by predicting future regional effects of various climatic variables and the study of various adaptation strategies. Diverse measures are necessary for reducing risks, like dry crops or improving watering efficiency (Ignaciuk and Mason-D'Croz, 2014). Appropriate and up-to-date corporate advice, education, and enhancement are essential for encouraging innovative technologies to facilitate the adaptive capacity of farmworkers (Ignaciuk, 2015).

Cities are distinctive in their capacity to meet the issue of global warming and climate change, but must be endorsed by regional and national policies, regulatory, and economic incentive schemes, and private funding. Cities are key to erasing their obstacles to adaptation to obtain financial assistance from other public administrations or the private industry (OECD, 2014a).

Climate change is impacting infrastructure in the energy sector, could interrupt energy supplies and changing requirements of energy demand. Building climate change resilience is among the major tasks for improving energy security, the International Energy Agency (IEA) considers.

Numerous methods are useful, such as risk assessment, innovative technologies, flexible management techniques adapted and emergency response indicators.

The IEA's annual Nexus Forum unites governments, business owners and other stakeholders in this sector to start sharing know-how and knowledge (IEA, 2014a). Several IEA flagships reports have integrated climate change adaptability in their assessment and key messages into the question, including World Energy Outlook. The nuclear energy industry is an instance of an energy sector which focuses greatly on climate change and the problem of adaptability.

Admittedly, the industry has become progressively aware of the effect that climate change, such as severe weather events, can have on the procedure and security of its crops given the long lifetime of nuclear power plants (40 to 60 years).

This has been part of the design of completely new plants and the regulatory framework, for example, in connection with the location of new plants that can be operated up to or beyond the 2080s. For current plants, safety regulations must be remedied so that protection does not jeopardize the level of risk based on external events, which include climatic events. The Nuclear Energy Agency (NEA, 2015) recommended that governments establish long-term adaptation investment structure and make sure that laws are amenable to adapting to climate change (Figure 9.5).

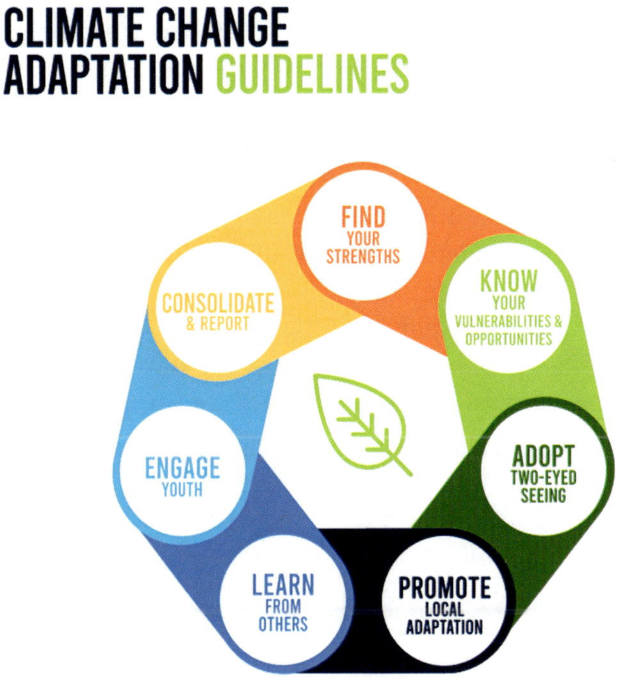

Figure 9.5: It is stressed that measures to raise versatility in climatic changes are needed to integrate adjustment in growth strategies and processes. This also shifts future development goals. Adaptation requires utilization of natural resources, encouraging food security, human, and social capital development, and reinforcement of organizational contexts.

Source: https://wikimab.org/climate-change-adaptation-guidelines-for-biosphere-reserves-in-canada-a-report-from-the-2019-climate-change-adaptation-forum/.

9.7. BENEFITS AND COSTS OF CLIMATE CHANGE

The Economic Impact of Climate Change (OECD, 2015b) examines the consequences for the various regions and sectors of the world economy of a variety of climatic impacts. Climate change is expected to have detrimental market repercussions in the overwhelming bulk of regions. Their activities involve agriculture (e.g., crop yield changes), coastal areas (e.g., capital, and sewage loss), some severe events (including hurricane-capital losses), wellbeing (extreme heat losses in labor productivity, disease expense, expenditure on health) and power and tourism requirement.

There are predicted to be substantial non-effects and dangers after certain climatic tipping points are attained. In Asia and Africa, where economic sectors are susceptible to a variety of climatic changes, like extreme heat and crop production losses, net economic effects are forecasted to be especially large. Macroeconomic costs are estimated to range from 1.5% to 6.5% of GDP in most nations in these areas by 2060 and average global cost of 1.0% to –3.3 of GDP. The formulated net financial benefits in Canada and Russia are forecasted in the next few decades at least to exceed the adverse effects. Climate adaptation and mitigation measures are both capable of reducing these effects and risks and reducing the macroeconomic costs resulting from the economy impacts chosen (Rothe, 2015).

Once GHGs are released into the atmosphere, the impact on climate and the economy will be inevitable for one or more 100 years, and the world will have higher consequences and poorer risk profile. In addition to reducing future costs of climate change and -more relevantly, reducing risks associated with increased temperature, optimistic policies of mitigation and adaptation may also lower the risk. Even so, despite the capability to mitigate effects in vulnerable areas, as in the many other countries in Africa and Asia, substantial effects of climate change are forecasted to remain. There will nevertheless be several market effects, even with ideal adaptation and mitigation laws in place (Hiltner, 2009).

Modeling the effects of climate change assists us in recognizing the comparative significance of impact modeling and their spatial distributions, through which impact societies. Even so, only a few of the prospective climate change expenses are included. Even the most elaborated designs entail only a subset of potentially changing climate opportunities and risks. The absence of proof for new, potentially devastating changes (e.g., destruction of the ice sheet), which rarely include these changes as a title, is one of the underlying difficulties.

Operation to counteract the effects and adapt to climate change is inspired, but also by the willingness to handle the effects that may occur. At national and regional levels, understanding is developing about climate risks and the subsequent benefits and costs of adaptation, but the media attention is irregular, and certain risks are still hard to model. The effect on coastal zones, coastal storms and floods have culturally been the most known climatic risks (Schuldt and Roh, 2014).

During the last decade, there has been a significant increase in the understanding on freshwater management, agriculture adjustment and construction. Ecosystem and biodiversity services, industry, and businesses face the most important data gaps remaining. Few government or private decision-makers have an overview of their changing climate exposure to risk. For instance, while some effects in government finances are routinely recorded, like disaster fund payments, there are not any other indirect costs like tax revenue effects. Moreover, governments are subjected to predicated obligations only when an event happens. Improved understanding of these effects and obligations can help countries control their climate risk exposure stronger. There will be no equitable burden of climate risks. Variables such as, gender health, revenue, and mobility impair the susceptibility and responsiveness of people. Even in countries with high average stages of economic development, differential impacts on vulnerable communities are a difficult task.

9.8. EVALUATION AND MONITORING

Given the magnitude of the adaptation obstacle, it is important to adopt and enforce appropriate techniques. Monitoring and review can enhance policy learning and enhance transparency by tracking the use of resources and the delivery of the legislation or project as anticipated. Four crucial instruments for adaptation monitoring and assessment are:

- Risk and vulnerability evaluations for climate change could provide a foundation for domestic climate change weaknesses, which can be evaluated in adaptation. Once again, the risk and vulnerability of such evaluations can also be demonstrated over time.
- Indicators promote an evaluation of advancement on adjustment objectives. Indicators cannot, however, themselves illustrate how the change occurred. Intensive resource reporting and use of indicators. Consequently, they need to be carefully characterized and draw on existing data sources if practicable.
- Assessments of projects and programs can help determine which adaptation strategies are effective in accomplishing agreed adaptation goals and understanding of some of their successful enabler factors (Schellnhuber, 2008).
- National climate audits and reviews analyze whether the allocation and cost-effectiveness of the resources allocated for adjustment is adequately targeted. This data can be helpful especially when resources for adjustment are explicitly allocated (Figure 9.6).

Figure 9.6: The possibility for natural adjustment can best be discovered by restricting the risk of harmful warming levels and by increasing aspirations and action for nature conservation, rehabilitation, and restoration. The climate change agenda fell at all levels of government during the COVID-19 pandemic. Convenience packages COVID-19 offer the chance to enhance the ambition of achieving climate and emissions resilient restoration.

Source: https://blogs.ei.columbia.edu/2021/01/21/report-climate-change-adaptation/.

The sophistication of the surveillance and assessment approach adopted must be directly proportional and the governance structure should not be under excessive influence. For example, new data may not be essential, because the ecological and socio-economic data already collected regularly by nations can notify adaptation monitoring and assessment. For instance, integrating relevant adaptation queries informed processes of data collection like household surveys can progressively address remaining data gaps.

Due to the variety of data used to measure and assess adjustment, a coordination mechanism can connect data creators and users with each other. For advancement cooperation providers in developing nations, they can encourage the advancement of the numerical systems of member states themselves through drawing on existing data collection processes for their own reporting requirements as far as probable (Strefler et al., 2014). When data deficiencies exist, providers of developed cooperation can implement programs which will contribute to the enhancement capacity in the data model of the partner country instead of focusing on collecting data for specific programs and projects. Monitoring and assessment are still in its early stages, and many OECD or partner countries still need to create structures for their national adaptation initiatives or methods.

Some of the countries that have taken up climate change adaptation include:

1. **Kenya:** Under the National Climate Change Action Plan in Kenya, the National Performance and Benefit Measurement Framework (2012) will track mitigation and adaptation activities. It integrates vulnerability indicators (bottom-up) and institutional capacity (top-down). They include both national (process-based) measures and county (outcome-based) measures and are strongly associated with the development plan for Kenya's Vision-2030.
2. **United Kingdom:** The National Adaptation Program Statutory Evaluation (2014) accepted for publication in 2015 utilizing comprehensive vulnerability indicators, realized effects and implementation of adaptive measures, along with assessment of decision-making procedures. The assessment framework was developed through sector-specific evaluations in the period 2012–2014 (water scarcity and flooding in the built environment, public health, business, infrastructure, and natural environment).
3. **Australia:** A new list of 12 indicators has been recognized as presently under discussion, and a National Adaptation Assessment Market is being developed. These metrics are organized around the economic and social drivers for adaptation to successfully adapt and the conclusions reached.
4. **Philippines:** Annually, the National Action Plan on Climate Change (2011) is measured and reviewed every 3 years, mostly using process-based measures, against instant, transitional, and results.
5. **Germany:** In September 2015, the Climate Change Adaptation Strategy (2008) Indicator system was authored. It consists of an assessment of weakness, an indicator-based assessment of 15 action and transversal areas of the strategy and a high-level evaluation as to whether the strategy is tackling the anticipated climate change opportunities and risks.
6. **France:** The 230 actions recognized in the French NAP have been established to represent several process-based indicators (2011–2015) (Figure 9.7).

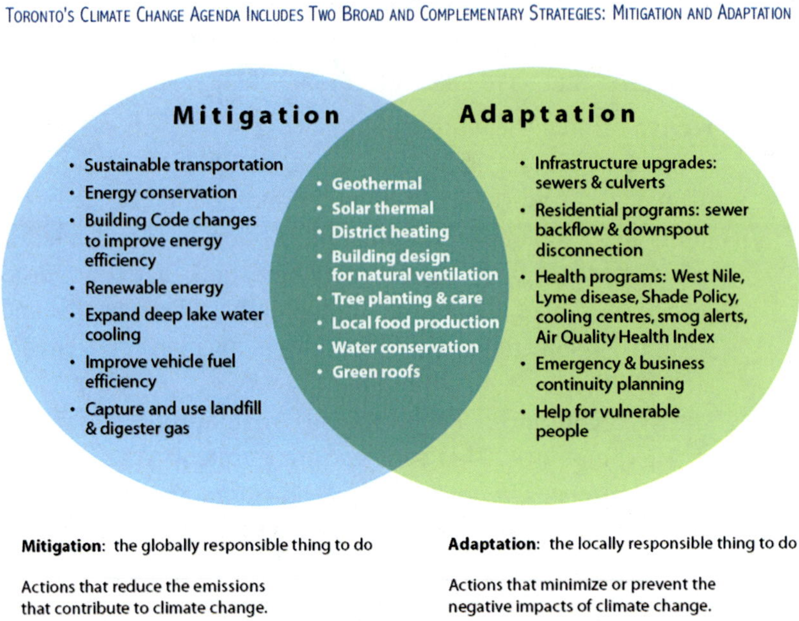

Figure 9.7: Many actions that encourage ecosystem adaptation also help people adapt to ecosystems. For example, restoring natural fire democracies decreases the possibility of catastrophic fires and decrease environmental exposure to this danger. Giving more room to the water bodies enables more water to be stored in the natural system and less probable to flood in populated areas. The supply of green areas and trees gives shade to animals. Trade is made with the reconstruction of ecosystems in certain areas between agricultural production.

Source: https://www.researchgate.net/figure/Green-spaces-as-means-to-both-mitigate-and-adapt-to-climate-change-4_fig1_228720496.

9.9. BARRIERS TO CLIMATE CHANGE ADAPTATION

As private and public actors participate in the execution of adaptation strategies, the level of adaptation preferred can be reduced by judgment obstacles. From a financial point of view, several obstacles, like transaction fees, discourage governments from making their adjustment decisions. Other obstacles arise due to market errors, like externalities, imbalances in data and moral risks. A full literature study recognizing the preceding financial barriers to decision-making adaptation is provided with the 5th Assessment Report of the International Committee on Climate Change (IPC CC) (2014):

1. **Cost of Transaction:** The data and adjustment costs are divided primarily. The first relates to costs in connection with knowledge dissemination and the second relates to the costs associated with long-term capital substitutions decisions (Summerhayes and Zalasiewicz, 2018).
2. **Mistakes in the Market and Markets Missing:** These include externalities, imbalances in data and ethical risks. This is especially the situation when maladaptation is considered when adaptation measures of one economic entity adversely affect and harm another unit (country, firm, individual, or sector). These market failures involve issues arising from an inadequate incentive system and thus give public officials a significant role.
3. **Adaptation Behavioral Barriers:** Contains all compartmental problems that lead, without utilizing all information available, to irrational decisions which are not time compatible. The effect on the adaptation judgment is also lacking in social norms and cultural variables.
4. **Distribution Concerns and Ethics:** The above-described vulnerability and adaptive volume distinctions are linked to these problems. Economic strategies normally recommend the choice of projects that are most cost-efficient, but ethical issues indicate that decision-making adaptation must address for net benefits as well as the effect on equity.
5. **Coordination, Failure of the Government, and Economic Policy:** Although public policymakers are assumed to erase the above obstacles, they suffer significant barriers, like inadequate knowledge or assets. In addition, a high level of cooperation between various levels of government requires most adaptation strategies.
6. **Incertitude:** This is one of the main challenges to adaptation, as it grows to various dimensions: future demographic, technological, and economic innovations, future climate change.

In conclusion, depending on the sensitivity and vulnerability to environmental effects, there is a different need for adaptation. In developing countries, adaptation is particularly important because it is most sensitive to climate change and has the biggest impact of global warming. Human adaptive capacity in various regions and populations is distributed evenly, and developing countries are usually less capable of adapting. The ability to adapt is intricately connected with economic and social development. Economic costs of climate change adaptation will probably cost billions of dollars annually over the next few decades, although there is no clear indication of the exact amount of money needed (Soutter and Mõttus, 2020). Governments as well as the active and ongoing commitment of stakeholders, including international, multilateral, regional, and national, the private and public sectors, civil society, and other stakeholders, and the efficient application of information are crucial for successful adjustments. Climate adaptation can be made across different regions, industries, and at different levels to the effects of climate change.

Chapter 10

How the EPA is Working with States and Communities to Limit Air Pollution and Climate Change?

CONTENTS

The EPA is an agency established in the United States and tasked with the duty of protecting the environment as well as human health. Their objective is to ensure that all three environmental aspects: land, air, and water are of good quality. They conduct frequent tests on the environment to check increase or decrease in pollution. They also establish laws to help curb the increase in environmental pollution. Apart from establishing laws, they also enforce environmental laws (Herndon and Whiteside, 2019). They also establish programs useful in cleaning up the environment. They also address issues dealing with environmental pollution, the most common being management of pollution. Different states have been working in conjunction with the EPA to control air pollution and climate change. This is as a result of the impacts of these variables on human and animal health.

10.1. COMBATING AIR POLLUTION

Air is a major component of the atmosphere and is very crucial in sustaining human life. The EPA is in charge of establishing air quality standards. They were first established in 1971. The EPA is required to release information on any updates. Information on air quality is very important. This is because over the years there has been a decline in visible air pollution. If the air is not studied, then there may be continuous release of poisonous gases to the atmosphere resulting in highly toxic air. EPA conducts frequent tests showing that the release of some pollutants in the environment is dangerous even at minimum quantities. To deal with air pollution, the EPA listed six major air pollutants required to meet air quality standards. These pollutants pose a significant threat to human health. Most states are working with the EPA to identify these pollutants and afterwards establish methods of controlling them.

10.2. CONDUCTING RESEARCH

The EPA has all the needed tools to conduct air research. Research is very crucial as it enables identification of pollutants and the extent to which they contaminate the air. Through results from research or studies on air quality, the state is able to make more informed decisions on managing air pollution. Research is done by the Office of Research and Development. The EPA has its own laboratories where they conduct investigation and studies on air samples. Most states help in funding some research. However, the EPA has its own ways of generating income. In some cases, the EPA may allocate some research to be done in academic institutions or in other capable organizations. These institutions need to meet the EPA's criteria for the research to be accepted. The office of Research and Development is given the mandate to select other organizations to be involved during research (Herndon, 2018).

There have been cases where the EPA has heavily relied on research from other areas. This is when its research department is not meeting its scientific needs. This is when the state may advocate for collaborative research. In a bid to gain more knowledge on air quality, the EPA has also established research programs. These programs allow studied to be down at ground level in different areas. The information is then collected and analyzed. Results are then released to relevant stakeholders to make decisions. With regards to research, the EPA makes use of technologically advanced instruments to monitor and measure the level of air quality. They also make use of highly advanced methods and techniques. Researchers continue to advance these tools to allow better determination of the level of air pollution. With very advanced technology, the EPA is able to support other research done on public health. Researchers can gain more knowledge on other sources of air pollution and how they are released in the atmosphere. They can also understand the extent to which different ecosystem are affected by air pollution. It allows easy classification of air pollutants. Availability of information allows the EPA to determine the management systems to be integrated in controlling air pollution (Hiltner, 2009).

Among the technologies used by the EPA include the New Fenceline Monitoring Technology. It is mostly used in industries. It measures the amount of air pollutants released in the atmosphere. It is a good technology as it is cost-effective. The technology also has the ability to control excessive production of air pollutants. The technology has been coupled up with the SPod sensor. This sensor has the ability to detect fumigative pollutants such as volatile organic compounds (VOC). The sensor is solar powered. It is put in an open space and is able to draw samples of air. Collected air samples are taken to the lab and are analyzed. EPA is also working on producing research tools allowing common scientists to conduct research in their home areas.

The EPA works with states to measure levels of air pollution in highly affected areas. A good example is long Island. The state and EPA had to put their heads together to protect long island. Air components such as carbon monoxide were measured. The study was aimed at identifying sources of air pollution. The study was also aimed at understanding the formation of ground ozone (An et al., 2018). They made use of the SPod technology in measuring carbon dioxide concentration in the atmosphere. Other chemical instruments were used in measuring concentrations of other chemicals in the environment. Release of aerosols in the air was measured using ceilometers. Ceilometers work by transmitting light impulses indicating the occurrence of a chemical reaction in the atmosphere. Spectrometers are used in monitoring movement of ground ozone from one area to another.

Once these pollutants are measured, researchers are able to measure the level in the atmosphere where these reactions take place. They are able to identify highly mobile pollutants. Some institutions have access to satellites. These satellites are fitted with sensors able to measure amounts of air toxins in the atmosphere. The EPA is looking into advancing aircraft studies to monitor air pollution in space and higher levels of the atmosphere. Results from numerous researches done by the EPA is used in making air quality standards. The state is required to comply with these standards.

10.3. STATE-EPA PARTNERSHIPS

The EPA collaborates with the state through partnerships. This is when the EPA is dealing with common pollutants. The protocol of dealing with common pollutants requires the establishment of national air quality standards. This is done by the EPA. Its main objective is to ensure the safety of individuals by drawing the line. Once established, the state has to play its part by developing plans seeking to implement national air quality standards. The state enforces these standards by ensuring all relevant bodies meet these standards. For example, in California, state officials work with local districts to develop strategies and plans to maintain air quality. Apart from producing plans, states need to come up with a list of compounds whose emission should be prohibited. Selection of these compounds should be done based on their impact on air quality.

The EPA is required to guide and help the state in making these decisions. If possible, representatives from the EPA should be involved in state planning. The state can make more informed decisions by utilizing EPA's issue on emission standards for newly discovered air pollutants. Once the state has made plains on air quality, a copy of the plan should be given to the EPA before its implementation. The EPA goes through the plans ensuring that they are in accordance with the air quality Act. Some considerations may be made on new sources of air pollutants. In most states, air agencies have the power to permit the implementation of air plans, while in other areas, the EPA is the only body permitting implementation on-air laws and plans (Al-Amin et al., 2013). When dealing with toxic air pollutants, a different approach is taken. The EPA shortlist toxic air pollutants. It then sets regulations on the emissions of these pollutants. The EPA has the power to prohibit and permit emission of these substances. In most cases, the EPA establishes limits on toxic air pollutants according to their sources. The limits may vary depending on the area of production and the source. For instance, in small areas, production of these pollutants may be minimal; hence they are not be severely affected by the limitations. Most of these limitations are heavy on industries as they are major producers. They will be required to reduce the production of toxic pollutants. After establishing the national air limits, the EPA requires the state to comply with the limits. However, the state has the power to choose whether they will adopt programs seeking to minimize the emission of toxic air pollutants. They can opt to partially adopt the program or fully adopt it. The state can come up with their own programs according to their needs and capability.

The following procedure is followed by the EPA when dealing with toxic air pollutants. It is a regulatory approach done in phases. The first phase is the setting of emission standards based on industry players. Research is done on industries already producing these pollutants. They look at the technology used by them. The EPA then sets limits based on availability of technology useful in minimizing production of these pollutants. In this phase, the EPA aim is timely toxic pollution management (Bassu et al., 2014). The same methodology is used when handling new sources. The goal is to ensure that large industries and new sources maintain a certain production line. They have to look for methods allowing them to copy with the regulations.

The second phase involves the EPA conducting a risk assessment. They check for potential health and environmental risks. This is under the residual risk program. After MACT standards have been established and implemented, the EPA has to wait for a certain duration before they can conduct a risk assessment. In most cases, the time frame is 8 years. They analyze health and environmental risks. If the level of production of these toxic pollutants poses health and environmental risks, then the EPA has to establish more limitations. This is to ensure human health is not at any risk and the environment does not suffer.

To better conduct risk assessments, the EPA has developed a risk assessment strategy. The strategy involves different methods being utilized depending on the source of the pollutant. They also make use of already existing tools used in determining the effects of exposing the surrounding to toxic air pollutants. They also borrow knowledge on different assessment methodologies from other organizations. Researchers can also develop their own methodologies. However, they need to be in accordance with the EPA's regulations (Hiller, 2011)

The most common method is the tiered approach. It entails numerous kinds of analysis being done. The complexity of the pollutant will determine whether the analysis should be more detailed. In most cases, screening analysis is done. Researchers make use of data on the pollutant. Already existing data has its own assumptions on these pollutants. The EPA conducts more studies on the pollutants, ensuring that all possible risks are taken into consideration. Comprehensive analysis requires more data on the pollutant. More resources will be needed for proper results. A demographic analysis is also crucial. It creates a better understanding on how these pollutants affect certain groups in the society.

When conducting a risk assessment, the technology used should possess all capabilities. Risk assessment should also be done on technologies involved in the production of these pollutants. The assessment should be done to determine whether new technologies have been developed to reduce the emission of pollutants. A technology review is also useful in identifying other measures put in place to minimize pollution. A technology review is optional. It is done based on the EPA's capability (Betts et al., 2011).

Standards have also been established to deal with acid rain. Acid rain is formed when rain combines with toxic substances present in the atmosphere. Congress from different states come up with federal programs dealing with acid rain. These programs seek to minimize the emission of gases forming acid rain. The EPA monitors the release of these substances in the atmosphere.

Another part of the environment under danger due to air pollution is ozone. The ozone layer is found at a certain level in the sky. This layer is useful in protecting human, animals, and the environment from harmful rays from the sun. Some of these Rays are carcinogenic. The EPA is in charge of ensuring the ozone layer is protected. They are establishing laws seeking to reduce the emission of gases and chemicals destroying the ozone layer. They also put in place a procedure to be used when handling these chemicals. Reduced production of ozone depleting substances to a certain degree prevents further damage to the ozone layer. The EPA has identified air pollutants destroying the ozone and has grouped them into two categories. The first group falls under class 1 while the second is in class 2.

Examples of substances in class 1 include chlorofluorocarbons (CFCs) and Halons. These substances can severely damage the ozone. After identification by the EPA, most countries have prohibited the production of these substances. Some countries have prohibited both importation and production of these substances. In other countries, some exceptions have been made. Hydrochlorofluorocarbons (HCFCs) constitutes substances in class 2. Most of them are used as substitutes for substances in class 1. Countries like the US have prohibited the production of new class 2 substances. Most class 2 substances are used in refrigerators and air conditioners.

The EPA has established some rules to be used for countries importing class 1 and 2 substances. There is a procedure to be followed when labeling and recording production of these substances. Government's need to monitor the production of these pollutants. For equipment using these substances, the manufacturer needs to give basic information on the equipment and follow the labeling protocol. Among the ozone pollutants, some considerations have been made with the production of some substances (Benjamin et al., 2017). They include methyl bromide and those used in laboratories and medicine production. The state, through members of Congress, is able to take necessary measures against the production of pollutants known to reduce visibility.

To better monitor the emission of pollutants, the EPA has established a law that requires all sources of pollutants to obtain permit. This applies for industries and other organizations. They first need to subject to federal air standards of the EPA before they get their permit. Local and state agencies handling air quality have the power to give these permits. The clean air act, given a guideline on who should receive a permit. Those who fall under major sources qualify. A major source that which has the ability to emission significant amount of pollutants exceeding the minimum threshold. It is also one which can produce about 100 tons of air pollutants in a year.

According to production of toxic pollutants, a major source produces an average of 10 tons in a year. Under the act, the following groups of people need permits: the municipal waste combustor, industrial, and commercial solid waste incinerators, and sewage sludge incinerators. Those under this category have to get permit regardless of the size of production. Those under non-major sources who need permits include secondary lead smelters, cement manufacturers, copper smelting industries, glass manufacturing industries, plants dealing with Mercury production, toxic waste combustors, and industries involved in gold production. Industries dealing with the production of chemicals fall under non-major sources and are required to obtain permits (Hiltner, 2009)

The following steps are followed when apply for a permit. Application forms are in the EPA's website. Information should be properly filled in the application form, after which it is summited to the EPA. Once the EPA receives an application, they send a representative who will evaluate the level to which the organization or industry complies with EPA's regulations. If they comply, the organization or industry will be awarded a permit for a certain duration. Once it expires, they need to renew their license. They also need to pay for the permit (Bose, 2010). When applying for the permit, there are fees to be paid at a certain frequency depending on the preference. There those who pay annually, monthly or after a number of years. Failure in payment results in fines.

The tribal government also has a part to play in ensuring reduced air pollution. Its role is to ensure that all laws under the Clean Air Act have been implemented. In some states, the tribal government, according to the desires of their tribes, can oversee a cleanup program. However, the tribe in question has to be approved by the EPA. If there are programs to be implemented by the tribe, the EPA should be aware and approve the project according to the air Act. By working with the EPA, the tribal government gets more knowledge on projects and programs approved by the EPA done at the community level. The tribal government is mostly found in India. There the EPA and members of the tribal government work together to protect the environment. They may also involve environmental specialists in developing air quality programs.

There are areas severely affected by air pollution. This is because the rate of air pollution in this area is relatively high. The EPA has shortlisted this arca and put in place some laws useful in helping these areas recover for pollution effects. Some of the measures taken up by the EPA include closing down of major sources in those areas. People living in those areas may be advised to take up alternative sources of energy. Other ways the EPA is working with states to minimize air pollution include identification of areas complying with air quality standards also those not comply with the standards (Buell, 2014). They also generate laws to help the state implement different air quality standards. The EPA is also conducting more research to help them support the implementation of certain air standards.

Areas not complying with the air standards may be required by the state to fully implement air quality plans. They also need to meet air standards established by the EPA failure to which necessary punishment can be imposed by the state on people living in those areas.

10.4. EPA AND COMMUNITY MEMBERS IN REDUCED POLLUTION

Members at the community level play a crucial role in minimizing pollution. This is because most activities causing pollution are done at the community level. Most industries producing air pollutants are found among people. Some activities such as burning of trees and other fuels are done at the community level. Major sources of air pollution such as cars are used by individuals. The EPA places much emphasis on the involvement of communities in combating air pollution. A major tool useful in controlling air pollution is education. One of the roles of the EPA is to educate the public on air pollution. The EPA educates the public on the causes and sources of air pollution. It also educates the public on ways they can help reduced air pollution. For instance, vehicle transmission is a major source of air pollutants. The EPA can advise members of the public to take up alternative means of transportation. They can advocate for the use of electric cars in place of cars running on petrol (Brulle et al., 2012).

The EPA also advocates for air pollution control projects done at the community level. In countries like India, there is a tribal government. This government helps members of their tribes identify projects. Once they have been identified, the tribal government gives the project proposal to the EPA for approval. The tribal government may find the project. If not, the EPA can allocate resources to be used during the project. Community members can also come up with projects to help reduce air pollution. All programs and projects need to be approved by the EPA before they are initiated.

Researchers in the EPA have developed numerous technologies allowing individuals to monitor air pollution. Among these technologies is the ATSDR Dose calculator. This program enables user to measure the amount of air toxins they have been exposed to hence the name exposure dose. The advantage of this tool is that it can measure other forms of pollution. The tool is user-friendly, allowing users to modify some parameters of measurement. There is also a set expected value allowing comparison.

Other than measuring technology, researchers have come up with other tools allowing local scientist to measure air pollution levels. They can then submit this information to the EPA. By using tools from the EPA, their results have higher chances of consideration (D'Amato et al., 2020). There some training tools given to members of the public. These tools enable individuals to familiarize themselves with technologies used in controlling air pollution. The training tool is detailed, giving information on a variety of equipment.

Risk assessment is known to be done by the EPA. However, the EPA has come up with projects aimed at helping the local public conduct their own risk assessment. These projects rely on a database provided by the EPA. From the database, individuals gain knowledge on different risk evaluations done at the community level for different areas. This information is useful in helping communities develop their own methodologies and assessments. They can also make use of information from previous assessments.

The EPA also deals with pollution issues by giving information on some activities causing air pollution. For instance, farming makes use of certain feeding operations which, when not properly done could cause air pollution. The EPA has a website. In the website, there is information on how feeding operations can be done without polluting the environment. There is also information on feeding regulations more so those related to air projects. With livestock, the issue is handling odors from animals. The website gives information on how farmers can handle these odors.

It is very important to ensure that all members of the public have information on the clean air Act. The need to be knowledgeable on what it entails and its requirements. They also need to be aware of the progress made by the state in improving air quality as well as limitations faced in controlling air pollution. All this information is provided in the website. In the site, there is information on how other bodies and interdisciplinary have been involved in controlling air pollution. There is an explanation of how the EPA develops and implements its laws. This way the public can bet understand laws madc and it makes it easier to comply with them (Jacobson, 2012)

The EPA works with the community in adopting better alternative sources of fuel and transport. Among them using proper school buses. This program is currently being done in the US. Old school buses running on Diesel and other fuels are being replaced with cleaner buses that emit less pollutants. As mentioned earlier, the EPA shortlist names of common pollutants and standards put in place to ensure controlled release of these pollutants. For the public to be knowledgeable, the EPA gives more information on these pollutants (Dai, 2011). They are able to identify sources, potential risks to human health, and environmental measures to minimize production.

On their website, the EPA outlines health indicators. This way, members of the public can make comparisons and know where they lie. They can also measure the extent to which pollution has reduced.

On their website, the EPA ranks counties according to their health status. Different variables are used in ranking. They include state of the environment, health tendencies, and provision of health care services. Most information obtained during community programs and projects are included in the EPA's database. This database can be accessed by organizations dealing with environmental care (Zhang, 2017). This data can be retrieved from numerous areas. Local leaders are also given guidelines on how they can improve air quality in their area. There is a leader's guide available on the EPA's website. It outlines different ideas useful in improving air quality. Children also need to be taught on the importance of controlling air pollution. The EPA provides tools for schools. This allows them to conduct studies on how to improve air quality. The toolkit contains a set of instructions useful to the learner. These are some of the ways the EPA is working with community members to improve air quality.

To help both the state and community members, the EPA gives grants. According to information provided on the EPA's website, there is a certain amount budgeted under grants. These grants are given to environmental programs and academic institutions, among others. Grant money is used in funding numerous environmental programs as well as scientific research. This results in cleanup programs in different areas. According to the EPA, giving of grants enables the meets their goal of safeguarding human and environmental health. It is estimated that the EPA uses about 4 billion as grants. Environmental organizations as well as nonprofit organizations benefit from grants given by the EPA. This is because they help the EPA actualize their goals.

Numerous positive impacts have been achieved due to grants given by the EPA. Allocation of grants is done by the EPA's office of grants. This office based on the criteria given by the EPA decides which organizations would be receiving grants. There are different types of grants given by the EPA. The first one is the multipurpose grants given to tribes and states. These grants are given to tribes and states to facilitate environmental programs. There grants given to student projects to facilitate studies and innovation. There are the environmental education grants (Diffenbaugh and Burke, 2019). This type of grants is used in facilitating public education on environmental programs this way, they can make more informed decisions. Air grants is used in facilitating projects formed to help improve air quality. It also deals with issues on climate change.

There are grants used in research institutions handling air contaminants. This allows them to purchase better tools allowing them to identify air contaminants. Some of the grants are used in developing online platforms where the EPA can share more information on air pollution. They can also exchange their information with other researchers. These funds are also used in attaining environmental justice. Relevant stakeholders work members of the EPA in creating solutions needed in solving environmental issues at different levels.

Grant is also used in campaigns against the use of fuels causing air pollution such as diesel. The grants fund projects in different parts of the country that campaign for reduced use of diesel fuels. They also develop programs aimed at reducing the emission of pollutants through engines. Some grants are allocated to students at the tertiary level, allowing them to come up with solutions to air quality problems. In some cases, the EPA calls for a stop in the production of certain substances because of the severe implications they pose on the environment. Prevention in the production of these elements is covered in pollution prevention programs and are funded by the EPA through grants.

With the growing use of technology, emphasis is placed on innovation. Young leaders are tasked with the duty of ensuring technology is adequately used in controlling pollution. With numerous areas to be supported through grants, the EPA has established some ways of getting income, among them includes fines charged on organizations not complying with EPA's regulations.

10.5. CLIMATE CHANGE

Among the numerous effects of climate change is air pollution. According to statistics released by the EPA in 2009, there is an expected increase in carbon dioxide production in the atmosphere, more so from Greenhouses. This will result in catastrophic effects on human health and the environment. Some environmental effects include frequent occurrence of acidic rain. It was also established that greenhouses established for numerous years produce long lived gases through the accumulation of these gases in greenhouses. These long-lived gases have severe impacts when released in the environment. Other than greenhouses, they are released from other human activities.

Research done by other organizations also points out the fact that climate change has mostly been brought about by human activities. They also indicate numerous risks brought about by climate change on the environment and human health. Some research list some of the effects of climate change on the environment, such as eutrophication, destruction of natural ecosystems. All studies done on the effects of climate change on health and the environment have consistent results (Donnelly et al., 2017). Risks posed by climate change on human health cannot be taken lightly. Recent research show that a large number of the population die due to disasters caused by extreme weather conditions. Others suffer from disease infections.

Trending Topics related to the environment include global warming. Global warming is caused by increased concentration of carbon dioxide in the atmosphere, which has led to the depletion of the ozone layer allowing most heat Ray's enter the Earth's atmosphere. Predictions made by scientist show that increased emission of carbon dioxide in the atmosphere could result in increased frequency of extreme weather events such as drought, storms, floods, hurricanes, and forest fires. This will result in numerous deaths and destruction of property.

Carbon dioxide and other gases released from Greenhouses are said to highly contaminate the air causing frequent heatwaves increasing numbers of individuals who die as a result of heat related illnesses. These cases are mostly recorded among children and the elderly. Climate change has raised a number of concerns with regards to public health. This is because there has been an increase in the production of ground ozone evident by the numbers of individuals hospitalized due to respiratory illnesses. Due to climate change, there has been an increased incidence of individuals suffering from vector-borne and water-borne diseases (Jang & Hart, 2015)

Other effects of air pollution include increased acidification of oceans, increase in sea levels, and strong storms. Increased acidification of oceans has caused species of aquatic animals into extinction. Other animal species suffer from a damaged ecosystem. Climate change has caused world governments to incur losses. It has also added to the pressure on food insecurity. This is because climate change has tampered with farming activities. As most of the changes in climate are brought about by air pollutants, the EPA has been actively involved in managing climate change (Fountain, 2019). This has involved numerous regulations being developed to limit the production of carbon dioxide, among other gases.

10.6. THE EPA AND THE STATE ON CLIMATE CHANGE

Among measures taken by the EPA and the state in managing climate change mostly revolves around reduced production of gases. This means controlling major sources. One of the major sources is exhaust from vehicles. The EPA is working with the traffic sector in managing pollution through exhaust. Their working together had resulted in the establishment of fuel standards for vehicles. These standards are applicable in light trucks and cars (An et al., 2018). There standards for medium-duty trucks as well as the heavy-duty trucks. This is because they have different rates of emission. Rules on fuels are established and implemented by the EPA. Their function is to ensure that fuels imported or produced by the state has a certain amount of renewable fuel.

Different projects have been initiated by the state and the EPA in fuel regulations. The EPA champions for the adoption of alternative sources of energy that are environment friendly. Among projects initiated by the EPA is the clean be projects. This project aims at replacing old buses running on diesel with buses running on other environment-friendly fuels. This results in a decline in the amount of exhaust released by school buses. The state and the EPA are involved in planning and implementing of these projects.

The EPA also issues a guide given to Congress. It helps them implement regulations on clean air. By so doing, programs such as the clean fuel fleet can be initiated. Another guide is the green vehicle. This give is mostly for the local public. It equips them with the necessary knowledge needed to purchase vehicles certified as environment-friendly. The guide contains information on technology used by smart card and the types of fuels they run on. There may be a comparison board showing the importance of using 'green fuels.' Since little progress has been made on changing fuels used by heavy-duty cars, the EPA gives a guideline on what the regulations require with the use of diesel fuels (Al-Amin et al., 2013).

The EPA also funds campaigns against the use of diesel fuels. They also fund acts seeking to reduce the emission of gases through exhaust. Funding is mainly through giving of grants. They are used in supporting projects and implementing laws on reduced diesel use. The EPA also form partnerships with companies offering environment friendly transport options. One company is the SmartWay company. They work together to create incentives for improved efficiency in managing gaseous emissions. The EPA is focused on providing information on different technologies used in controlling diesel use. The Diesel Retrofit technology is among them. By availing this technology to the public, more research can be done on fuels.

The EPA is working with the state through non-governmental organizations. These organizations are working towards a healthier environment. These organizations use forums such as the Diesel Technology Forum in creating awareness on the use of diesel fuels and engines. They educate the public on some modifications made to improve fuel efficiency and minimize fuel emission. They also track progress made by the EPA, state, and organizations in minimizing air pollution. In the US, there is the green fleets organization. They are concerned with the introduction of alternative sources of fuel. They have their own initiative managed by an international council involved with the local environment.

The EPA and the state are working on producing permits for greenhouse owners. The permits are given before construction begins. These permits list the minimum threshold for the production of gases from major sources. Refining industries, steel mills, and power plants are some of the major sources required to have permits. Permits can be given at two stages, when they are under construction, meaning production has not begun. They can also be given when the plans are undergoing huge modifications. Permits given for major sources are termed as tittle V. Programs have been initiated to help with issuing oh permits though in most cases, environmental agencies have been given this mandate (Herndon and Whiteside, 2019). Many programs have been developed because the clean air Act requires the presence of numerous programs to make issuing faster and easier. This enables the EPA to monitor emission in several areas.

The clean air Act regulations may seem too severe for a state. In this car, the EPA can tailor a rule to help the state better adjust to regulations. Some of the tailored rules have made some consideration on production of certain gases. They have been given a new threshold which changes as production continues. They are also given permits on the basis of the amount of gaseous pollutants released in the atmosphere. The final part of the tailored rule is bent on allowing GHG reduce limits placed on major sources. This allows them to retain their initial threshold. This allows the EPA to implement reasonable limitations based on the type of emission.

Various state is recognizing efforts put in place by the EPA in controlling climate change. In 2015, former President Obama made an announcement on plans with the EPA to produce clean power. Researchers argued that it was the best way to go as it would reduce excessive emission of carbon in the environment. Production of clean power means that power plants will have to make use of other environment-friendly options during production. This will result in reduced changes in climate.

Some of the advantages of using clean air include the fact that it allows flexibility, reduced environmental pollution and its fair form both power plants and the environment. Establishing clean power plants is expected to have fast growth. It will be picked up by other countries, making it a global trend. These power plants can be modified to meet the state's power demand. With the help of the EPA, states can achieve environmental goals. The same applies for the EPA (Bassu et al., 2014).

According to statistics, the use of clean energy will significantly cut carbon dioxide production, causing a reduction in global warming and other extreme weather events. Decisions by government in adopting other sources of energy show that they are committed to working with the EPA to ensure that air pollution and climate change is managed. Use of clean power may require old power plants to make some adjustments to be in line with new power plants. The transition may be quite expensive however, the adoption of clean energy has proven to be reliable and affordable once it has been set up.

The clean air Act allows the EPA to collaborate or form partnerships with the community, tribe, and the state for its goals to be achieved. Depending on the partnership formed, the EPA and its partners can develop new goals as well as a strategic plan. This will help them be more oriented towards achieving their goals. The procedure followed during formation of partnerships is listed under the clean power plan. A lot of research is done before the clean power plan is implemented. The EPA is required to issue new air quality standards for newly established power plants. They also need to present proposals on federal plans and models useful in achieving successful implementation.

The Supreme Court has also been involved in implementing clean power plans. Before implementation, the plan is given to members of the judiciary who will then review it before they can implement it. The court has to look into two different sources of information weighing the merits and demerits of using the plan. They should not be swayed by the beliefs of the EPA. When the plan was presented before the court, the EPA firmly believed that the plan would be implemented as it was formed on the basis of strong scientific foundation. The court ruling did not meet the EPA's expectation. This resulted in a repeal in 2017. (Zaval et al., 2014).

Other than the clean power plan, the EPA is working to implement strategies aimed at reducing the emission of methane. An issue of the strategy was released in March 2014. They have also made clear their plans to reduce methane production from the oil sector. Their phased plan includes a 45% in methane production by the year 2025. Some of these plans are being implemented in the USA. This can be termed as a very ambitious goal considering the extensive uses of methane. Later on, the EPA made proposals seeking to implement measures to cut production and emission of methane (Herndon, 2018). This will result in reduced smog formation. Provision of laws governing cut in production enables those in industries have a sense of direction. They will also have an idea of the path to be taken by the oil and gas industry. There were proposals on areas having high methane production, such as the solid wastelands governed by the municipal office. The proposal involved reduced release of gases rich in methane. This resulted in the development of programs allowing industries to measure the release of these gases. They also measure compliance of these industries in reducing pollution.

As hydrofluorocarbons (HFCs) are used in the production of methane, the EPA has prohibited the use of these substances. Plans are underway to make this rule effective in other states. This calls for the use of alternative source of HFCs in the different areas of use.

10.7. HANDLING CLIMATE CHANGE AT THE COMMUNITY LEVEL

One method used by the EPA in managing climate change is educating the public on climate change. They provide information on the causes of climate change and solutions to problems caused by climate change. They can also educate the public on ways they can prevent climate change. This involves engaging community members in projects and programs such as clean school bus project. They can also encourage the public to come up with their own programs. Educating the public also involves educating young children as they are the future generation. They are taught of alternative sources of energy and how they can best preserve the environment.

In learning institutions, more so at the tertiary level, the EPA fund research projects. This facilitates more studies on climate change expounding on already existing data. Young students are also encouraged to be innovative and come up with transportation mechanism safe for the environment. They can also develop new technologies useful in monitoring the emission of gases in the atmosphere. They can also develop technology to be used in monitoring and controlling wildfires. This may include heat sensors and automated fire controllers (Singh & Purohit, 2014).

Climate change is best handled at the community level through good community planning. Statistics show that vehicle emissions are the leading cause of climate change at the community level. This is because there are many individuals who use vehicles on a daily basis. This form of emission is mostly in urban areas. There is significant emission of carbon through energy use. Fuels are burnt to produce energy used in cooking, lighting, and cooling. The fact that most individuals are dependent on cars makes the situation worse. The EPA works with community members by enlightening them on alternative sources of energy. For instance, rather than using vehicles to work, they may use bicycles, and if the distance to work is not long, then they can walk to work (Betts et al., 2011).

The EPA also encourage them to use alternative sources of fuel. They discourage poor farming methods and operations that could contribute to carbon dioxide emission. Practices like burning of land to clear vegetation has been prohibited in some countries. Burning of forests or trees to produce charcoal has also been prohibited. As trees help in maintaining a natural air cycle, community have been actively involved in planting trees. Afforestation and re-afforestation programs have been initiated to increase tree coverage. These projects have also been initiated in school for children to take up such practices.

For air pollution and climate change to be properly managed, there is need for collaboration between the state, community members and the EPA. When the state follows regulations listen in the clean air Act, a significant decline will be noted in the rate of air pollution (Benjamin et al., 2017). The state can also help in funding projects initiated by the EPA. It requires every individual to be their brother's keeper and ensure they use environment resources properly so that future generations enjoy an environment free from pollution.

Bibliography

1. Hiller, A., (2011). Climate change and individual responsibility. *The Monist, 94*(3), 349–368.
2. Hiltner, K., (2009). Renaissance literature and our contemporary attitude toward global warming. *Interdisciplinary Studies in Literature and Environment, 16*(3), 429–441.
3. Jacobson, M. Z., (2012). *Air Pollution and Global Warming: History, Science, and Solutions*. Cambridge University Press.
4. Jang, S. M., & Hart, P. S., (2015). Polarized frames on "climate change" and "global warming" across countries and states: Evidence from twitter big data. *Global Environmental Change, 32*, 11–17.
5. Kellstedt, P. M., Zahran, S., & Vedlitz, A., (2008). Personal efficacy, the information environment, and attitudes toward global warming and climate change in the United States. *Risk Analysis: An International Journal, 28*(1), 113–126.
6. Kenis, A., & Mathijs, E., (2012). Beyond individual behavior change: The role of power, knowledge, and strategy in tackling climate change. *Environmental Education Research, 18*(1), 45–65.
7. Kgabi, N., Grant, C., & Antoine, J., (2014). Effects of energy production and consumption on air pollution and global warming. *Journal of Power and Energy Engineering, 2014*.

8. Kump, L. R., (2011). The last great global warming. *Scientific American, 305*(1), 56–61.
9. Lineman, M., Do, Y., Kim, J. Y., & Joo, G. J., (2015). Talking about climate change and global warming. *PloS One, 10*(9), e0138996.
10. Mika, J., Forgo, P., Lakatos, L., Olah, A. B., Rapi, S., & Utasi, Z., (2018). Impact of 1.5 K global warming on urban air pollution and heat island with outlook on human health effects. *Current Opinion in Environmental Sustainability, 30*, 151–159.
11. Milfont, T. L., (2012). The interplay between knowledge, perceived efficacy, and concern about global warming and climate change: A one-year longitudinal study. *Risk Analysis: An International Journal, 32*(6), 1003–1020.
12. Peter, S. C., (2018). Reduction of CO_2 to chemicals and fuels: A solution to global warming and energy crisis. *ACS Energy Letters, 3*(7), 1557–1561.
13. Philander, S. G., (2008). *Encyclopedia of Global Warming and Climate Change: AE* (Vol. 1). Sage.
14. Pleijel, H., (2009). *Air Pollution and Climate Change: Two Sides of the Same Coin?* Swedish Environmental Protection Agency.
15. Revesz, R. L., Howard, P. H., Arrow, K., Goulder, L. H., Kopp, R. E., Livermore, M. A., & Sterner, T., (2014). Global warming: Improve economic models of climate change. *Nature News, 508*(7495), 173.
16. Robinson, R. A., Crick, H. Q., Learmonth, J. A., Maclean, I. M., Thomas, C. D., Bairlein, F., & Visser, M. E., (2009). Traveling through a warming world: Climate change and migratory species. *Endangered Species Research, 7*(2), 87–99.
17. Rothe, D., (2015). *Securitizing Global Warming: A Climate of Complexity*. Routledge.
18. Schellnhuber, H. J., (2008). Global warming: Stop worrying, start panicking? *Proceedings of the National Academy of Sciences, 105*(38), 14239, 14240.
19. Schuldt, J. P., & Roh, S., (2014). Media frames and cognitive accessibility: What do "global warming" and "climate change" evoke in partisan minds? *Environmental Communication, 8*(4), 529–548.
20. Shapiro, M. A., & Park, H. W., (2015). More than entertainment: YouTube and public responses to the science of global warming and climate change. *Social Science Information, 54*(1), 115–145.

21. Shepardson, D. P., Niyogi, D., Choi, S., & Charusombat, U., (2011). Students' conceptions about the greenhouse effect, global warming, and climate change. *Climatic Change, 104*(3), 481–507.

22. Singh, A., & Purohit, B. M., (2014). Public health impacts of global warming and climate change. *Peace Review, 26*(1), 112–120.

23. Soutter, A. R. B., & Mõttus, R., (2020). "Global warming" versus "climate change": A replication on the association between political self-identification, question wording, and environmental beliefs. *Journal of Environmental Psychology, 69*, 101413.

24. Strefler, J., Luderer, G., Kriegler, E., & Meinshausen, M., (2014). Can air pollutant controls change global warming? *Environmental Science and Policy, 41*, 33–43.

25. Summerhayes, C. P., & Zalasiewicz, J., (2018). Global warming and the Anthropocene. *Geology Today*, *34*(5), 194–200.

26. Trenberth, K. E., Dai, A., Van, D. S. G., Jones, P. D., Barichivich, J., Briffa, K. R., & Sheffield, J., (2014). Global warming and changes in drought. *Nature Climate Change, 4*(1), 17–22.

27. Victor, D. G., (2011). *The Collapse of the Kyoto Protocol and the Struggle to Slow Global Warming*. Princeton University Press.

28. Villar, A., & Krosnick, J. A., (2011). Global warming vs. climate change, taxes vs. prices: Does word choice matter? *Climatic Change, 105*(1), 1–12.

29. Wang, Y., (2015). Air pollution or global warming: Attribution of extreme precipitation changes in eastern China—comments on "trends of extreme precipitation in Eastern China and their possible causes." *Advances in Atmospheric Sciences, 32*(10), 1444–1446.

30. Yeager, D. S., Larson, S. B., Krosnick, J. A., & Tompson, T., (2011). Measuring Americans' issue priorities: A new version of the most important problem question reveals more concern about global warming and the environment. *Public Opinion Quarterly, 75*(1), 125–138.

31. Zaval, L., Keenan, E. A., Johnson, E. J., & Weber, E. U., (2014). How warm days increase belief in global warming. *Nature Climate Change, 4*(2), 143–147.

32. Zhang, R., (2017). Warming boosts air pollution. *Nature Climate Change, 7*(4), 238, 239.

INDEX

S

T

U

V

W